THE FAIRLIE LOCOMOTIVE

David & Charles Locomotive Monographs

General Editor:
O. S. Nock, BSc, CEng, FICE, FIMechE

Published titles
The Caledonian Dunalastairs, by O. S. Nock
The GWR Stars, Castles & Kings, Parts 1 and 2, by O. S. Nock
The LNWR Precursor Family, by O. S. Nock
The Midland Compounds, by O. S. Nock
The Stirling Singles of the Great Northern Railway, by Kenneth H. Leech and Maurice Boddy

In preparation
The Gresley Pacifics, by O. S. Nock
The Standard Gauge 4-4-0s of the GWR, by O. S. Nock

Locomotive Studies

The Fairlie Locomotive, by Rowland A. S. Abbott
The Garratt Locomotive, by A. E. Durrant
The Steam Locomotives of Eastern Europe, by A. E. Durrant
Steam Locomotives in Industry, by the Industrial Locomotive Society

Robert Francis Fairlie, 1831-85

DAVID & CHARLES LOCOMOTIVE STUDIES

THE FAIRLIE LOCOMOTIVE

ROWLAND A. S. ABBOTT

DAVID & CHARLES
NEWTON ABBOT

ISBN 0 7153 4902 3

COPYRIGHT NOTICE

© ROWLAND A. S. ABBOTT 1970

All rights reserved. No part of this publication may be reproduced, stored in a retrieval system, or transmitted, in any form or by any means, electronic, mechanical, photocopying, recording or otherwise, without the prior permission of David & Charles (Publishers) Limited

Set in 10-point Plantin
and printed in Great Britain
by W J Holman Limited Dawlish
for David & Charles (Publishers) Limited
South Devon House Newton Abbot Devon

CONTENTS

			PAGE
Preface			9
Chapter	1	Introduction	11
	2	The Experimental Period, 1865-1870	13
	3	Great Southern & Western Railway, Ireland, Inchicore Works, 1869	22
	4	Sharp, Stewart & Co Ltd, Manchester, 1870-79	24
	5	Avonside Engine Company Bristol, 1871-81	27
	6	The Yorkshire Engine Co Ltd, Sheffield, 1872-1906	43
	7	The Vulcan Foundry Limited, Newton-le-Willows, 1872-1911	49
	8	R. & W. Hawthorne & Company, Newcastle upon Tyne, 1874-90	57
	9	Wiener-Neustadter Lokomotivfabrik vormals G. Sigl, 1879	60
	10	Festiniog Railway, Boston Lodge Works, 1879-85	62
	11	Neilson & Company, Glasgow, 1883-1901	65
	12	The Kolomensky Works, Russia, 1884	67
	13	Sächsische Maschinenfabrik vormals Richard Hartman, Chemnitz, Germany, 1902	69
	14	North British Locomotive Co Ltd, Glasgow, 1903-08	71
	15	The Hunslet Engine Co Ltd, Leeds, 1908	73
	16	Miscellanea	75
	17	Commentary	77
	18	The Mason–Fairlie Locomotive, 1871-1914	80
	19	The Péchot–Bourdon Locomotive, 1888-1921	92
Bibliography			96
Acknowledgments			97
Sources of photographs and drawings			98
Index			99

PREFACE

THE nucleus of this book originally appeared as a series of articles by the present writer in *The Engineer* during 1958 and 1960, and these articles rapidly proved that there existed a widespread interest in this unusual design of articulated steam locomotive.

The resulting correspondence with locomotive engineers and historians in many parts of the world has brought to light much additional information, and enabled a number of errors to be corrected, while at the same time many extra illustrations have become available. Although it is of such unique interest, it is very remarkable that in all the mass of literature on the steam locomotive no previous book has ever been published dealing exclusively with the history of the Fairlie locomotive, while references in the technical press over the last hundred years are also very meagre and confined to incidental references, or detached articles on a particular class (usually a well known one); yet the Fairlie type was constructed over a period of some forty-six years and formed a valuable contribution to the motive power of many overseas railways where heavy loads were operated over difficult terrain.

In preparing the text no concession has been made to prosy dialogue, indeed a technico-historical work of this kind does not lend itself to such treatment, and it is considered that this will be appreciated by the serious student of the subject, to whom concise details and illustrations are of prime importance, and it is hoped therefore that this book will prove of interest to both engineers and historians and fill a niche in the literature of the steam locomotive.

R. A. S. ABBOTT

Old Botley, Oxford

CHAPTER 1

INTRODUCTION

A SMALL number of unsuccessful locomotives designed on the principle of a double boiler mounted on two power-bogies were tried many years before R. F. Fairlie made it a practical success. These were Horatio Allen's three 2–2–2–2 locomotives built at the West Point Foundry, New York, in 1831; the 0–4–4–0 Semmering Contest locomotive *Seraing* built at the Cockerill Works at Seraing, Belgium, in 1851; and the Thouvenot design, French Patent No 59,773 of 1863.

Robert Francis Fairlie was born in Scotland in March 1831 of an engineering family, and received his training as a locomotive engineer at Crewe and Swindon. In 1853 he was appointed to the position of Engineer and General Manager of the Londonderry & Coleraine Railway, and afterwards went to India on the Bombay & Baroda Railway. He married, in 1862, the daughter of George England, of the Hatcham Ironworks, and by the early 1860s he was established as a railway consultant with offices at 56 Gracechurch Street, London, E.C., and was concerned with railways then under construction in South America and India.

In 1864 Fairlie published a 36-page pamphlet, with 13 illustrations, entitled 'Locomotive Engines' in which he expounded, by means of a dialogue between an engineer and a writer, the supposed shortcomings of steam locomotives of conventional design. He then proceeded to show how his newly conceived ideas for an articulated locomotive would overcome all objections, by the superior steam generating capacity of the twin boilers, the use of all wheels for adhesion, the ability to traverse sharp curves, the abolition of turntables, etc.

There is some indication (pages 6 and 33) that he may have been more influenced by Archibald Sturrock's locomotives with steam tenders, and (page 11) by the back-to-back tank locomotives built by R. Stephenson & Co in 1856 for the Great Indian Peninsular Railway Ghat inclines, than by any previous knowledge of the *Seraing* 0–4–4–0 of 1851. Fairlie appears to have had a great appreciation of

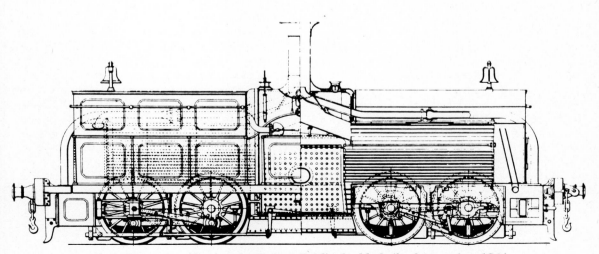

Fig 1 *Patent specification drawing of Fairlie double-boiler locomotive, 1864*

THE FAIRLIE LOCOMOTIVE

the value of publicity in business and loudly proclaimed the virtues of his new locomotive, with the result that over the years he has been regarded rather as a propagandist, but upon reading the above-mentioned pamphlet the fact emerges that his thinking was in many ways ahead of his time.

Robert Fairlie took out a British Patent, No 1210 of 12 May 1864, and a French Patent, dated 23 November 1864, for 'Improvements in Locomotive Engines and Boilers', involving a new system of articulated locomotive which differed in many respects from the Seraing locomotive. The patent specification extends to fourteen pages, with three large folding plates of twenty-one drawings, but it will be sufficient for the present purpose to reproduce here only the general arrangement drawing (Fig 1) which, however, exhibits the classic features that later became associated with all Fairlie locomotives, together with some unusual details that were never used in practice.

In addition to the double boiler with central firebox and two smokeboxes, the patent also covered the possible use of an upper and lower series of tubes in these boiler barrels, the first series leading from the firebox to each smokebox in the normal way, while the upper series were to have led the gases back to a smoke chamber situated over the firebox; this chamber being in communication with a central chimney. No Fairlie locomotives were ever constructed with these return tubes and central chimney. Although the double boiler was one of the main features of the original patent, it was not an essential one, and Fairlie himself prepared designs for single-boiler tank locomotives with one power-bogie and a normal trailing bogie, and a design for a proposed locomotive with a single boiler and two power-bogies. The double-boiler locomotive, with two power-bogies, was a design from which, by the early 1870s, a great performance was expected, and after the famous trials on the narrow-gauge Festiniog Railway with the *Little Wonder* in 1870, the Fairlie principle was widely acclaimed and a big demand for these locomotives was anticipated from overseas railways. In practice, however, the type was found to possess several disadvantages, among these being the limited quantity of fuel and water that could be carried; also repairs and maintenance of a double boiler are naturally more costly than for a single one, although this disadvantage is compensated to a certain extent by the fact that the tube heating surface is more efficient than in an ordinary locomotive boiler having the equivalent heating surface. For example, two sets of tubes, each 8 ft 6 in long, are more efficient than one set 17 ft long, with the same heating surface.

Lastly the type lacks stability at high speeds and this limited its use to lines where such speeds were not required. On steeply graded railways with sharp curves the design achieved a certain degree of success, and under such conditions a considerable number of these locomotives continued to give satisfactory service for many years. But quite a number of the early Fairlies often failed to live up to their first magnified prestige, probably due to poor maintenance facilities overseas, and there were recurring troubles with the original designs of articulated steam pipes, and so the engines were withdrawn from service after only a few years' work.

Between 1865 and 1911 the classic Fairlie locomotive was built by ten British and three Continental firms, and in two railway workshops for England, Wales, Ireland, Canada, Colorado, Mexico, South America, Cuba, New Zealand, Australia, Cape Colony, India, Burma, Norway, Sweden, France, Luxembourg, Saxony, Portugal, and Russia.

The two most successful applications of the Fairlie principle were undoubtedly on the Trans-Caucasian railway system in Russia, where forty-five such locomotives were used, chiefly on the Souram inclines, and on the Mexican Railway, on the main line of which, between Cordova and Bocca del Monte (108 miles) no other form of steam locomotive was used over a period of fifty-three years, until the Arizoba–Esperanza section of the line was electrified in 1924.

For obvious reasons, the Fairlie locomotive was never seriously considered by the railways of Britain, there being neither the gradients nor severe curves to warrant extended use of such a heavy and specialised design, and the eight railways which did acquire examples were all, with one exception, minor companies.

Only for the unusual operating conditions existing on the narrow-gauge Festiniog Railway in North Wales has the Fairlie principle justified its use in this country.

CHAPTER 2

THE EXPERIMENTAL PERIOD
1865-1870

IN a little over a year from the granting of the patent the building of the first double-boiler locomotive, with two power-bogies, was undertaken by Messrs James Cross & Company, Sutton Engine Works, St Helen's, Lancashire, and was completed in December 1865. This firm occupied the workshops of the former St Helen's Railway, and constructed some sixty locomotives between 1864 and 1869. This locomotive was the *Progress* for the Neath & Brecon Railway and it attracted considerable attention, extended references to it being published in the engineering journals of the time. The boiler consisted of two barrels, with a common firebox in the centre, and under each barrel was an engine bogie, each of which formed a separate power unit. The bogies, the pivots of which were riveted to the undersides of the barrels, had four coupled wheels, 4 ft 6 in diameter, the distance between the axles being 5 ft. The cylinders were outside, 15 in diameter by 22 in stroke, and placed at the outer or smokebox ends of the bogies. The boiler barrels had a diameter of 4 ft and each was provided with 191 brass tubes, 2 in diameter outside and 9 ft 6 in long, providing a heating surface of 950·5 sq ft.

The firebox was constructed generally to accepted practice, but naturally was provided with a tubeplate at either end and a firedoor in one side. It had a grate 6 ft 6 in long by 3 ft 10 in wide, giving an area of almost 25 sq ft. There was no water partition and the draught from one chimney had a tendency to draw air down the chimney at the other end. Fairlie stated that this error was due to the builders,

Fig 2 *The* Progress; *the first Fairlie locomotive built, by Cross & Co in 1865 for the Neath & Brecon Railway*

THE FAIRLIE LOCOMOTIVE

who apparently decided that a water partition would be a disadvantage. The firebox heating surface was 92 sq ft, which, together with that of the two boiler barrels, made up a total heating surface of 1,993 sq ft. The steam pressure was 120 psi. A footplate was provided on either side of the firebox, the firedoor being on the opposite side to the driver's position where the regulator handle and reversing lever were located. The bunkers, extending alongside each boiler on the fireman's side, had a capacity of about 2 tons 5 cwt of coal, and the four side tanks had a combined capacity of almost 2,000 gallons. The total weight of the locomotive in working order was 42 tons. The braking system was unusual, the brake blocks being applied, by chain gearing operated from the fireman's side of the footplate, to friction wheels fitted on the axles. The single dome on the firebox top had a spring-balance safety valve on either side. To provide for the movement of the bogies, each main steam pipe was of copper and coiled once round inside the smokebox, passing out to the steam-chest through a slotted aperture at the bottom of the smokebox. This was the obvious weak point in the design, and there was constant loss of steam between the boiler and the steam-chests due to the difficulty of keeping the flanged joints tight. The coiled steam pipes eventually failed when negotiating a sharp curve, and a steam pipe of the pendulum pattern was afterwards fitted.

A photograph of this historic locomotive is reproduced in Fig 2. In 1869 the *Progress* was sent to the Midland Railway, and in May and June of that year worked between Kentish Town and Hendon; later in the same year it is thought to have been purchased by George England & Company of the Hatcham Ironworks, Pomeroy Street, New Cross, London. George England and Robert Fairlie were at this time associated with the development of the latter's patent locomotive, and the *Progress* may have been acquired for experimental or demonstration purposes. The ultimate fate of this locomotive is uncertain, but it was reported to have been tried on the Talybont incline of the Brecon & Merthyr Railway in June 1870. It was not purchased by the latter company, and was sold in July 1870 to the Monmouthshire Railway. In June 1871 it was put up for sale by auction at Newport and after this nothing further seems to be known about it.

A smaller, but generally similar locomotive was built for the Anglesey Central Railway and delivered by Cross of St Helen's in August 1866. This second Fairlie locomotive (Fig 3) was named *Mountaineer*, and except for the brake gear, the provision of ball-and-socket joints on the steam pipes, and the difference in dimensions, the description already given for the *Progress* also applies to this locomotive. The brake screw on the fireman's footplate worked, through a vertical bell crank lever, a horizontal lever working on a pin fixed under the ashpan. This horizontal lever was connected to a rod attached to

Fig 3 *The* Mountaineer *built by Cross & Co in 1866 for the Anglesey Central Railway*

THE EXPERIMENTAL PERIOD

the centre of a transverse plate, at either end of which were wooden brake blocks acting, on each bogie, on the pair of wheels nearest the firebox. The diameter of the wheels was 4 ft 10 in, the wheelbase of each bogie 4 ft 11 in, and the total wheelbase 19 ft 11 in. The cylinders were 10 in diameter by 16 in stroke. Each boiler barrel was carried on a forging riveted to the underside, and the lower side of each forging formed a pin 5 in in diameter. In addition to the bogie frames there was a connecting frame round the firebox casing. The boiler barrels measured 2 ft 9 in in diameter and were 9 ft 4 in long, each barrel having 99 tubes of $1\frac{5}{8}$ in diameter. The firebox casing was 5 ft long by 3 ft 5 in wide, the inside box being 4 ft $4\frac{1}{2}$ in long and 2 ft 10 in wide; grate area 12·4 sq ft. The total heating surface amounted to 858 sq ft, being made up of 402 sq ft for each set of tubes and 54 sq ft in the firebox, which had a water space partition. The bunker was rather small and only 15 cwt of coal could be carried, while the combined capacity of the four side tanks was 900 gallons. The total weight in working order was 34 tons.

The Anglesey Central Railway was opened throughout from Gaerwen Junction to Amlwch on 3 June 1867, and Mr Hanbury Miers, who was a director of this line, was also the first Chairman of the Neath & Brecon Railway, so this fact may account for the transfer of the *Mountaineer*, after only a few months' work, from the Anglesey Central Railway to the Neath & Brecon Railway. According to the late Glen A. Taylor (1880–1935) of Neath, this locomotive had a working life of thirteen years, after which it was sold to the Briton Ferry Iron Works and dismantled, the boiler being sent to their Llantrisant Tinplate Works. At a later date the wheels were sold, but not the axles, while one cylinder, complete with guide bars and connecting rod, was fitted with a crankshaft and sent up to the Neath works of the company for driving the fitting shop, where it remained in use until 1896. The boiler was eventually returned from Llantrisant to the Briton Ferry Works and there cut up.

It will be observed that the above two locomotives had only one steam dome, and this was centrally mounted on the firebox. The provision of a single dome was not a feature of later double-boiler Fairlie locomotives, and after about 1870 it became standard practice to mount a dome on the back ring of each boiler barrel.

It is therefore of some interest to recall that the Péchot-Bourdon double-boiler locomotives, which differed but little from normal Fairlies, had one large dome on the firebox, and in the patent specification of 3 June 1887, one of the improvements claimed was 'a central position for the steam dome ensuring that the steam shall always be taken at a constant height above the water level of the boiler'.

In 1866 three 0–6–6–0 locomotives were built by Cross & Co (Works Nos 28, 29 and 30) for the Southern & Western Railway of Queensland, Australia. These locomotives (Fig 4) were built to the designs of Charles Douglas Fox, CE, and had cylinders 11 in diameter by 18 in stroke, wheels of 3 ft diameter, bogie wheelbase 6 ft 6 in, distance between bogie centres 16 ft 10 in, total wheelbase 22 ft 6 in. The rail gauge was 3 ft 6 in. The main bearing springs were of the transverse kind, placed one below each axle. As springs so arranged have no lateral stability, each bogie was provided, in addition, with four volute springs—a pair near each end—these being placed between brackets fixed to the inside of the main frames and brackets on the boiler barrels. The transverse distance between these springs was only 2 ft 1 in and it is doubtful whether they proved very effective.

The boiler barrels had a length of 9 ft 10 in and a diameter of 3 ft, each containing 133 tubes of $1\frac{1}{2}$ in diameter. The tube heating surface amounted to 1,037·4 sq ft, and that of the firebox 72·1 sq ft, making a total heating surface of 1,109·5 sq ft. The grate area was 17·4 sq ft. The central dome contained a regulator of the grid-iron type, and from it the steam passed into a pipe formed by riveting a bent plate along the inside of the boiler top. From this 'pipe' the steam entered two small steam-chests placed one on each barrel, and from these it was conducted by branch pipes to other chambers fixed on the bottom of each barrel, these chambers being annular and forming the sockets into which the bogie pins fitted. From each annular chamber a forked pipe led to the cylinders, this pipe being connected to the chamber by a stuffing-box which allowed the pipe to follow the movements of the bogie. The blast pipes were also arranged so that they could follow the movements of the bogies and at the same time deliver their jets in the centre of the chimneys. The eccentrics were placed in a very inconvenient position, outside the frames but behind the driving wheel bosses; the valve gear being of the shifting link pattern, with communication to each slide valve through a rocking shaft.

The feed water was carried partly in tanks formed by using the sides of the bogie frames together with the necessary plates and angles, and partly in the side tanks situated underneath the footplate on each side of the firebox casing, and on the footplate on each side at the bottom of the coal boxes. The com-

THE FAIRLIE LOCOMOTIVE

bined contents of all these tanks amounted to 1,030 gallons. The four coal boxes had a capacity of 120 cubic ft. The total weight of these locomotives was 30 tons. These three locomotives were designed to handle loads of 90 tons up a gradient of 1 in 45 which extended for a distance of 12 miles, and with curves of 330 ft radius.

As was customary at the period when these three locomotives were built, they were shipped in a dismantled condition, and on arrival in Australia two were erected and put to work under test conditions, but they were never a success and their very poor performance did much to retard the acceptance of the Fairlie locomotive as a practical machine. All three were eventually returned to England and sent to the works of the Yorkshire Engine Co at Sheffield, where they were rebuilt to the 4 ft 8½ in gauge, with drastic alterations including the moving of the valve gear inside the frames and discarding the original water tanks. One, Works No 29, was then sold to the Burry Port & Gwendraeth Valley Railway in 1873, and on this line it carried the name *Victoria*; it is illustrated as rebuilt in Fig 5.

When working on this South Wales railway the wheels are recorded as being 2 ft 9 in in diameter, the capacity of the side tanks as 1,000 gallons and the total weight as 37 tons empty.

In 1896 this locomotive was rebuilt in the company's workshops with two new boiler barrels of Lowmoor iron and a new copper firebox; the working pressure was then 140 psi.

In 1899 it was numbered 8 and finally scrapped at Burry Port in 1903. The other two locomotives were sold to the Central Uruguay Railway in 1874, but proved of very little use except on slow goods trains. One was withdrawn in 1881 and the other in 1896.

The return of these three locomotives from Australia and their subsequent re-sale was reported in *Engineering* for 2 March 1877. In September 1869, Robert Fairlie, in conjunction with George England (Junior) and J. S. Fraser (of the Great Western Railway), acquired the Hatcham Ironworks of George England & Co at New Cross, London, and under the title of the Fairlie Engine & Steam Carriage Company proceeded to specialise in the construction of the patent double-boiler locomotive.

The first to be completed, in 1869, was the well-

Fig 4 *Locomotive built by Cross & Co in 1866 for the*

THE EXPERIMENTAL PERIOD

known *Little Wonder* for the 1 ft 11½ in gauge Festiniog Railway in North Wales. This locomotive (Fig 6) had cylinders 8 3/16 in diameter by 13 in stroke with valve gear of the Gooch stationary link type. A bogie of 5 ft wheelbase was placed under each boiler barrel with coupled wheels of 2 ft 4 in diameter, each bogie frame being pivoted in a plain bearing on the centre line of the coupled wheelbase. The distance between the bogie centre was 14 ft 1 in and the total wheelbase 19 ft 1 in. The boiler barrels were each 7 ft 6 in long by 2 ft 6 in diameter, each containing 109 tubes of 1½ in diameter and 7 ft 10 in long. The main outside firebox casing measured 6 ft in length by 3 ft wide and curved inwards at the sides with two internal fireboxes each 2 ft 5¾ in long by 2 ft 7 in wide and 3 ft 6 in deep with their adjacent ends made semi-circular. The heating surface of the 218 tubes was 670 sq ft, while the fireboxes provided 60 sq ft, making a total of 730 sq ft. The combined grate area of the two fireboxes amounted to 11 sq ft and the working pressure was 160 psi, the tractive effort being 5,375 lb.

Salter safety valves were mounted on the two steam domes, each of which contained a horizontal slide type regulator worked by a common handwheel on top of the firebox casing. Although having the appearance of side tanks, the water tanks were box-like containers which completely enveloped the tops of the boiler barrels, the combined water capacity being 680 gallons. One tank carried a large wooden toolbox along its top, and the coal was carried on the top of each tank in front of the weatherboards; the total bunker capacity was only 15 cwt.

The locomotive had dome covers of polished brass and bell-mouthed chimneys with copper caps, while an unusual feature was the provision of small snowplough-like guards in front of each bogie.

This Festiniog example was the first narrow-gauge application of the Fairlie patent, and a series of trials was undertaken with this locomotive during the year 1870 in the presence of many eminent engineers when, on account of its performance, the design was widely acclaimed and the technical press of the day was full of the potentialities of the Fairlie principle.

The first cost of the *Little Wonder* was £1,950, and although it was apparently so successful this

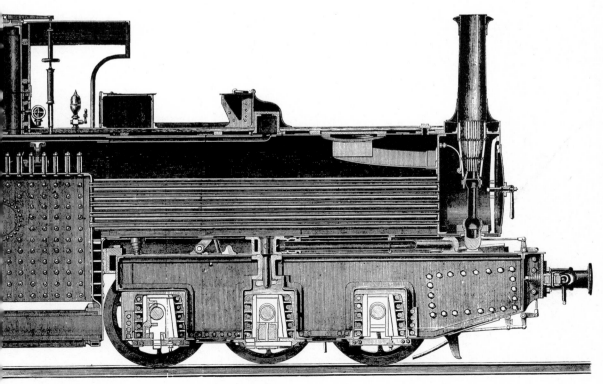

Southern & Western Railway of Queensland. 3 ft 6 in gauge

THE FAIRLIE LOCOMOTIVE

locomotive was never rebuilt, being withdrawn in 1879, condemned in 1882, and broken up in 1883. Two locomotives of the 0–4–4–0 type were ordered by the Nassjo-Oscarshamn Railway in Sweden, but the first of these, when completed in December 1869, was sold instead to the Burry Port & Gwendraeth Valley Railway in South Wales. This locomotive, originally named *Pioneer*, was renamed *Mountaineer* and is illustrated by Fig 7. The photograph shows a name plate on the edge of the footplate, the inscription reading 'FAIRLIE ENGINE AND STEAM CARRIAGE CO. LONDON. 1870'. In 1877 it was tried on the severe gradients of the Pantyffynon to Rhos line of the Great Western Railway. The length of time that this locomotive remained in service is not known, but it appears to have been dismantled at Burry Port before 1900; the bogies being used for the transport of machinery to the local collieries, while the boiler shells were converted for use as a culvert at West Dock, Burry Port. The second of the locomotives ordered for Sweden was delivered there to the British contractor, J. Morton & Sons, and was named *Morton*. After the failure of this firm it was transferred to their successors, Clark, Punchard & Co who renamed it *Clark*. Later it carried the No 1 of the railway, and when taken over by NOJ in 1874 it was renamed again, *Hultenhein*. It ceased work in 1902 after a boiler explosion.

A third locomotive, identical with the above pair, was built for the Chemins de fer de la Vendée in France and was officially recorded as being in use there in 1871. When this line was taken over by the State system in 1879 this locomotive was already out of use.

These three small locomotives had the following leading dimensions: the cylinders were 10 in diameter by 18 in stroke; wheels 3 ft 6 in diameter, with unusual divided balance weights; bogie wheelbase 5 ft; total wheelbase 19 ft 6 in. The boiler barrels were 8 ft 6 in long by 2 ft 10 in diameter. Heating surface of tubes 763 sq ft, fireboxes 70 sq ft, making a total heating surface of 833 sq ft. Grate area 11 sq ft. Capacity of water tanks 700 gallons, while the coal bunkers held 15 cwt. The total weight was 25 tons. It should be noted that no brake blocks are visible in the photograph; was some form of band brake used?

In 1870 a 0–6–6–0 locomotive, the *Tarapaca* (Fig 8), was completed for the Iquique Railway in Peru (due to territorial changes in this area the city

Fig 5 *Southern & Western Railway of Queensland 0–6–6–0 locomotive as rebuilt in 1873 by the Yorkshire Engine Co, sold to the Burry Port & Gwendraeth Railway, and named* Victoria

THE EXPERIMENTAL PERIOD

Fig 6 *The* Little Wonder *of 1869, built by the Fairlie Engine & Steam Carriage Co for the Festiniog Railway. 1 ft 11½ in gauge*

Fig 7 *Burry Port & Gwendraeth Valley Railway; the* Mountaineer *built by the Fairlie Engine & Steam Carriage Co in 1869*

THE FAIRLIE LOCOMOTIVE

of Iquique is now in Chile), a line owned at that time by Messrs Montero Brothers and which was later to become part of the Nitrate Railways. Intended for a duty that involved the hauling of 150 tons up an incline of 1 in 26 for a distance of 11 miles, this locomotive had cylinders 15 in diameter by 20 in stroke; wheels 3 ft 6 in diameter; wheelbase of bogies 7 ft 8 in; distance between centres of bogie pins 20 ft 3 in; total wheelbase 27 ft 11 in. The length of boiler barrels was 10 ft 6 in by 3 ft 10 in diameter. The heating surface of the tubes amounted to 1,500 sq ft, and that of the two internal fireboxes to 125 sq ft, giving a total heating surface of 1,625 sq ft. The grate area was 21 sq ft. The large water tanks held 2,200 gallons. The valve gear was Allan's straight link motion, and the eccentrics had cranked rods in order to pass above and below the central axle of each bogie. The boiler was carried by transverse bearers of suitable form which extended under each barrel near the middle of its length and connected the pair of 'carrier' frames which passed along the sides of the firebox casing at a higher level than the bogie frames. These carrier frames served to connect the two bogie pins and to transmit any pull or thrust from one bogie to the other, so relieving the boiler barrels of all traction stresses. The bogie pivots each had a flat bearing surface 1 ft 8 in in diameter with a hole $6\frac{1}{4}$ in diameter in the centre. The bogie pin, of gunmetal, passed through this hole and was made hollow in order to conduct steam to an annular chamber from which a steam pipe led to the cylinders.

At the rear of each bogie, next to the firebox, the frames were connected by a cast-iron transverse stay, the upper flange of which had a slot curved to an arc struck from the centre of the bogie pin. Through this slot passed a pin which served to connect the bogie frame with a plate iron bracket on the underside of the boiler, there being placed on this pin a number of india-rubber rings which served to steady the bogie frame and check any tendency to 'kick' when the locomotive was pulling hard. This arrangement was covered by the original patent specification, and with various modifications was used on all future Fairlie locomotives.

When the *Tarapaca* was tried on the 'circle' at Hatcham Ironworks it traversed the curves of 50 ft radius with ease; these curves were, of course, quite

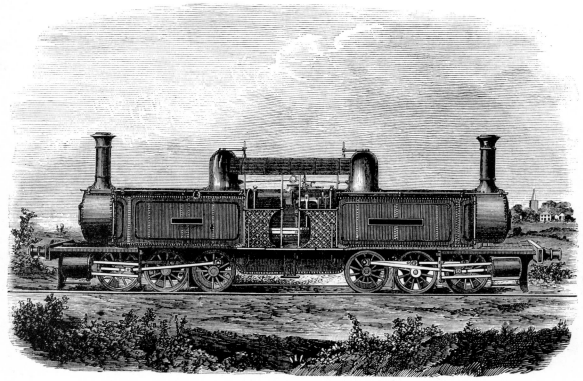

Fig 8 *The* Tarapaca *of the Iquique Railway, Peru. Built in 1870 by the Fairlie Engine & Steam Carriage Co*

THE EXPERIMENTAL PERIOD

exceptional and on the railway for which it had been built the sharpest curves had a radius of 800 ft. With a weight of 60 tons this locomotive was, at the time of its construction, one of the most powerful in the world. The Hatcham Ironworks does not seem to have prospered under the new management, for the above five locomotives were the total output up to the time that the business was closed down and the machinery sold by auction on 14 May 1872.

However, under the title of the Fairlie Engine & Rolling Stock Co offices were taken at Palace Chambers, Victoria Street, Westminster, and locomotives continued to be designed from 1872 until 1878 for construction by outside firms.

An article by Robert Fairlie with the title of 'Die Alpenlokomotive der Zukunft' ('The Alpine Locomotive of the Future') was published in *Technische Mitteilungen* at Zürich during 1876, and from this it is apparent that by March of that year his patent locomotive was in use on forty-three railways; the list actually enumerates forty-seven lines, but four of these were branches of one system.

In the course of a professional journey in 1873 to Venezuela, Fairlie contracted blood poisoning and sunstroke, a misfortune from which he never fully recovered, although he continued to take an interest in his business affairs until the day before his death which occurred on 31 July 1885.

From 1870 onwards several of the old-established locomotive building firms took up the manufacture of the Fairlie double-boiler type, and in view of this it will be convenient to deal with the productions from these builders in the chronological order in which the initial locomotive appeared from each works, but before proceeding further something may be said at this point on the subject of the so-called hybrid-Fairlies which were introduced as early as 1869. These combined the features of a normal 0-4-4 or 0-6-4 tank locomotive with bunker, tanks and a single boiler of conventional form, on one frame, supported upon a pivot, with a very small amount of traverse, in the usual manner at the rear, but with the cylinders and driving wheels mounted as a bogie, with a pivot supporting the boiler. Only a small number were built by British firms, and in view of this they will be described under builders along with the normal Fairlies and not dealt with under a separate heading.

It may be mentioned here that it was William Mason in America who developed this design to its greatest extent, and who built 148 such locomotives of eight different wheel arrangements between 1871 and 1889 at his works at Taunton, Massachusetts, although the final examples were built as late as 1914 by the American Locomotive Co. These American locomotives are described in Chapter 18.

CHAPTER 3

GREAT SOUTHERN & WESTERN RAILWAY, IRELAND, INCHICORE WORKS, 1869

THE first example of the single-boiler Fairlie type locomotive was designed by Alexander McDonnell (1829-1904) and built at the Inchicore Works of the Great Southern & Western Railway, Ireland, where Mr McDonnell was locomotive superintendent from 1864 until 1882. This locomotive (Fig 9) was built in 1869 and numbered 33, and was of the 0–4–4 back-tank pattern with front coupled wheels and inside cylinders forming a steam bogie and a hind bogie of the Adams type with lateral movement and india-rubber pad; the two bogies being connected by a carrying frame supporting the boiler, bunker and tank. The pivot of the steam bogie had a flat face, which rested in a recessed socket and although nearly all the weight was carried on the pivot there were also india-rubber load checking springs on each side in a cup bracket attached to each main carrying frame.

This arrangement allowed the steam bogie and carrying frame to tilt together laterally, whilst the trailing Adams bogie could tilt independently of the main frames. Any tendency to pitch on the part of the steam bogie was checked by india-rubber springs fixed to a bracket carried in the centre of the transverse stay in front of the firebox. Compensating levers connected the springs of the coupled wheels and the trailing bogie had a single inverted spring on either side. The steam pipe of 3 in diameter was taken out of the front of the dome into a tee-piece, from which two external pipes passed round the boiler barrel to a tee-joint underneath the centre line of the locomotive. From this point a single bent pipe conveyed steam to the valve chest between the cylinders. There was no swivelling joint as on modern articulated locomotives and the movement of the steam bogie was accommodated only by the elasticity of the pipes and bends. The blast pipe was fitted with a petticoat pipe, the lower mouth of which was large enough to allow the bottom portion of the pipe to move within it. The cylinders were 15 in diameter by 20 in stroke; the total heating surface was 738 sq ft, and the grate area 14·6 sq ft. The coupled wheels, 5 ft 7½ in in diameter, had a wheelbase of 6 ft, the pivot of the steam bogie being 2 ft 7 in in front of the crank axle. The trailing bogie, the wheels of which had a diameter of 2 ft 11 in, had a wheelbase of 5 ft. The distance between the bogie centres was 14 ft 7 in; the total wheelbase 20 ft 6 in; weight on steam bogie 22 tons; total weight 35 tons 17 cwt. The water tank held 800 gallons and the bunker 30 cwt of coal. A side elevation drawing is shown in Fig 9a. This locomotive was found to run with great steadiness and was able to pass round curves of 300 ft radius with ease—indeed it was possible to negotiate curves of 200 ft radius, but the driving wheels then almost touched the carrying frame. A second locomotive of this design, No 34, was completed at Inchicore Works in 1870.

During a discussion at the Institute of Civil Engineers in 1873, Mr McDonnell stated that up to that year No 33 had run 30,977 miles and No 34 had run 22,738 miles; the coal consumption had averaged 19·8 and 21 lb per mile respectively. The average train consisted of 6 six-wheeled carriages.

It is not known whether the steam pipes gave any trouble, but both locomotives remained at work for more than twenty years. They were broken up about 1892.

From two surviving letters, dated 4 and 16 September 1869, written by Fairlie to C. E. Spooner of the Festiniog Railway, it appears that he visited Ireland in September of that year to see the first of these single-Fairlie type locomotives at work.

GREAT SOUTHERN & WESTERN RAILWAY, IRELAND

Fig 9 Great Southern & Western Railway of Ireland, No 33. The first example of the single-boiler Fairlie type locomotive. Built at Inchicore Works in 1869. 5 ft 3 in gauge

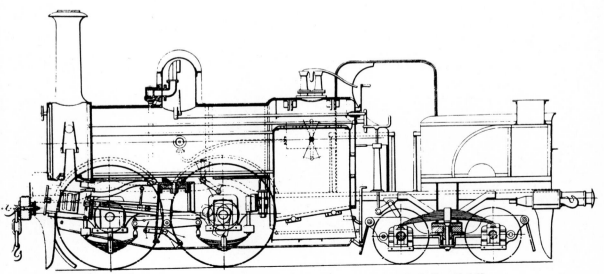

Fig 9a General arrangement drawing of the first single-boiler Fairlie as built at the Inchicore Works in 1869

CHAPTER 4

SHARP, STEWART & CO LTD
ATLAS WORKS, MANCHESTER, 1870-79

In 1870 two 0-4-4-0 locomotives were delivered by this firm to the Nassjo-Oscarshamn Railway, Sweden (Works Nos 2190 and 2191), followed during 1871 by two more (Works Nos 2239 and 2240). These four locomotives had cylinders 10 in in diameter by 18 in stroke; wheels 3 ft 6 in diameter; tube heating surface 763 sq ft; firebox 70 sq ft; total 833 sq ft. The bunkers held 50 cu ft, and the capacity of the water tanks was 900 gallons.

This class was very similar to the *Morton*, and their railway numbers, names and subsequent history were as follows:

2	*Eskjo*	re-boiled 1897	scrapped in 1911
3	*Nassjo*	re-boilered 1894	scrapped in 1911
4	*Oscarshamn*	re-boilered 1896	scrapped in 1910
5	*Hultsfred*	. . . . .	sold 1899 to a contractor

During February 1871 an Imperial Russian Commission paid a visit of inspection to the celebrated Festiniog Railway and there witnessed some interesting trials with the locomotive *Little Wonder*, and as a result of this a recommendation was made by the commission for the building of a system of narrow-gauge lines in Russia.

In May 1870 construction had commenced on the Imperial Linvy Railway, the gauge adopted being 3 ft 6 in, and the line, which was a little over 38 miles long, was completed in April 1871. To work the new line an order for five 0-6-6-0 locomotives was placed with the Atlas Works (Works Nos 2088-2092) and these were delivered in 1871.

These locomotives (Fig 10) had cylinders 13 in in diameter by 18 in stroke; wheels 3 ft 3 in diameter; bogie wheelbase 7 ft 6 in; bogie centres 20 ft; total wheelbase 27 ft 6 in. The boiler barrels were 10 ft 6 in long by 3 ft $3\frac{1}{2}$ in diameter, having a tube heating surface of 1,210 sq ft, which, together with 115 sq ft provided by the two internal fireboxes, gave a total heating surface of 1,325 sq ft. The

Fig 10 *Imperial Linvy Railway, Russia. Built in 1871 by Sharp, Stewart & Co. 3 ft 6 in gauge*

SHARP, STEWART & CO LTD

Fig 11 *Locomotive for Tamboff & Saratoff Railway, Russia. Built in 1871 by Sharp, Stewart & Co. 5 ft gauge*

grate area was 21 sq ft. The boiler pressure was 140 psi, and the estimated tractive effort was 15,600 lb. The water tanks held 1,250 gallons and the fuel racks could accommodate 200 cu ft of wood, while the total weight of the locomotive in working order was 46 tons.

One detail in which these Linvy locomotives differed from those previously built was in the arrangement of the centre pins of the bogies. The frames of each bogie were braced together by transverse and diagonal stays near the middle of their length, and these transverse stays supported a cast-steel socket in which brass centre pins rested. As in all later Fairlie locomotives, the boiler was supported by carrier frames to which the bogie pins were fixed. These centre pins were 12 in in diameter and made with flat ends, but the whole weight of the boiler did not rest on these pins, each of the latter being supplemented by two segmental blocks which rested on a plate carried by the above-mentioned transverse stays. These locomotives were followed in 1871 by ten of the 0-6-6-0 type for the Tamboff & Saratoff Railway (Fig 11).

This railway commenced about 275 miles southeast of Moscow and ran about 200 miles to the banks of the Volga.

Built to the Russian gauge of 5 ft, these locomotives (Works Nos 2118-2127) had cylinders 15 in in diameter by 20 in stroke; wheels 3 ft 6 in diameter; bogie wheelbase 8 ft; total wheelbase 29 ft 3 in. The boiler barrels had a length of 11 ft 3 in and a diameter of 2 ft $9\frac{3}{4}$ in, the combined tube heating surface being 1,500 sq ft. With 125 sq ft contributed by the two fireboxes the total heating surface was 1,625 sq ft. The grate area was 21·3 sq ft. The capacity of the tanks was 1,800 gallons and that of the bunkers 300 cu ft of wood. The weight empty was 45 tons 18 cwt. These Tamboff & Saratoff locomotives were all sold to the Poti & Tiflis Railway in 1887. The last Fairlie locomotives built by Sharp, Stewart were two of the 0-4-4-0 type (Works Nos 2867 and 2868) supplied to the order of the Brazilian Government for the 3 ft $7\frac{5}{8}$ in gauge Canta-Gallo Railway; the actual contract being placed by Phipps & Co as agents.

The principal dimensions were: cylinders $10\frac{1}{2}$ in diameter by 18 in stroke; wheels 3 ft 3 in diameter on a bogie wheelbase of 5 ft, the total wheelbase being 19 ft 7 in; boiler barrels 8 ft 6 in in length, with an inside diameter of 2 ft $9\frac{1}{4}$ in; 102 tubes of $1\frac{5}{8}$ in diameter in each barrel; tube heating surface 758·4 sq ft; fireboxes 73 sq ft; total 831·4 sq ft. Each firebox had a width of 2 ft $6\frac{3}{4}$ in and was 2 ft $2\frac{1}{2}$ in long with a depth from crown to firebars of 3 ft 11 in. The centre-line of the boilers above rail level was 5 ft $5\frac{1}{2}$ in, and the overall length of the loco-

THE FAIRLIE LOCOMOTIVE

motive, excluding buffing and draw gear, was 28 ft 5 in. The water tanks held 800 gallons. The design is illustrated by Fig 12.

This Brazilian line is the one, now of metre gauge and part of the Leopoldina Railway, that runs from Niterói (across the harbour from Rio) to Nova Friburgo, about fifty miles. The unusual gauge of this railway may be due to the fact that the section between Boca do Mato and Teodoro de Olivera, a distance of ten miles, was built for working on the 'Fell' centre-rail system using material and rolling stock from the 3 ft $7\frac{5}{8}$ in Mont Cenis Railway, which was dismantled in 1871. The date of conversion to metre gauge is not known.

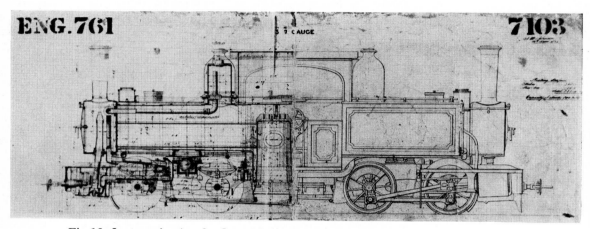

Fig 12 *Locomotive for the Canta-Gallo Railway, Brazil. Built to the unusual gauge of 3 ft $7\frac{5}{16}$ in by Sharp, Stewart & Co in 1873*

CHAPTER 5

AVONSIDE ENGINE COMPANY BRISTOL, 1871-81

THE Avonside Engine Company took up the manufacture of the Fairlie locomotive in 1871, and between that year and 1881 produced more of this type than any other British builder. It is therefore all the more regrettable that the records of this firm, now in the possession of the Hunslet Engine Co Ltd, Leeds, are very incomplete and only five official photographs of Fairlies are known, while the general arrangement drawings of only six designs now survive.

It should be noted that it was this firm's custom to allocate two works numbers to each double-boiler locomotive, although this was not always strictly adhered to in practice.

The earliest locomotives were for the nitrate area in Peru; the first (Works Nos 853-4) built in 1871 (Fig 13), is mentioned in *Organ für die Fortschritte des Eisenbahnwesens*, 1872, page 174, where it is stated that 'The locomotive *Hercules* has been built in Bristol for the Iquique Railway in Peru and is of the same kind as another locomotive which has been built before for the same company'. It was a sister locomotive to the *Tarapaca* from Hatcham. This was followed by three other 0–6–6–0 locomotives for the Pisagua Railway in Peru (Works Nos 851-2 of 1871 and 903-906 of 1871-2). All these had cylinders 15 in diameter by 22 in stroke; wheels 3 ft 6 in diameter; total heating surface 1,685 sq ft;

Fig 13 *The* Hercules *built by the Avonside Engine Co in 1871 for the Iquique Railway, Peru*

THE FAIRLIE LOCOMOTIVE

coal capacity 80 cu ft; water 2,200 gallons; weight 60 tons.

The Pisagua Railway was a line running from the Pacific coast eastward; some fifty miles south was the Iquique Railway, also running inland from the coast. Joining the two lines at their inland termini was the Nitrate Railway running north and south. The Nitrate Railway absorbed the Iquique in 1871 and the Pisagua a year or so later. The whole system became known as the Nitrate Railways in 1882, with headquarters and workshops at the city of Iquique. For the next few years the three names were used rather indiscriminately, causing some concern to later-day historians. A map of these railways was published in *The Locomotive* for 1932, page 85. The nitrate area was formerly in the republic of Peru, but was ceded to Chile in the 1880s. The 3 ft 6 in-gauge Toronto, Grey & Bruce Railway in Canada ordered one 0–6–6–0 in 1871 (Works Nos 862-3) and this locomotive, the *Caledon*, is illustrated in Fig 14. The cylinders had a diameter of $11\frac{1}{2}$ in by 18 in stroke; wheels were 3 ft 3 in diameter; bogie wheelbase 7 ft 6 in; total wheelbase 26 ft 3 in. The boiler barrels had a length of 10 ft and were 3 ft $0\frac{3}{4}$ in diameter; the tube heating surface being 858 sq ft. The heating surface of the two fireboxes was 103 sq ft, and the grate area amounted to 20 sq ft. The water tanks held 1,400 gallons, and 200 cu ft of wood could be carried. This locomotive was scrapped in 1881 when the TG & B was made a standard-gauge line under Canadian Pacific control.

An identical locomotive (Works Nos 864-5) was also supplied in 1871 to the neighbouring Toronto & Nipissing Railway. This 3 ft 6 in-gauge line was the first narrow-gauge railway to be opened for traffic on the continent of America; this was in 1869.

The Poti & Tiflis Railway in Georgia was commenced in 1869 as a 5 ft-gauge line from Poti, a small port on the Black Sea and completed to Tiflis, a distance of 169 miles, in October 1872. Near the summit of the Souram Pass the gradient is 1 in $22\frac{1}{2}$ and to work this difficult section four 0–6–6–0 locomotives were built in 1871 (Works Nos 872-879). These locomotives (Fig 15) had cylinders 15 in diameter by 22 in stroke; wheels 3 ft 7 in diameter; total heating surface 1,685 sq ft; water capacity 2,200 gallons, and they were originally woodburners. It was customary to use three of these 60 ton machines on passenger trains over the pass. The line was later extended to link Batoum on the Black Sea with Baku on the Caspian Sea, and later became State owned, being known as the Transcaucasian.

During 1871 three 0–6–6–0 locomotives were delivered to the Ferrocarril Mexicano—The Mexican Railway Co Ltd. These locomotives (Works Nos 880-885) had cylinders 15 in diameter by 22 in stroke, and wheels of 3 ft 7 in diameter. The railway numbers were 21 to 23. A fine photograph of this class, in original condition, was published on page 16 of the American journal *Trains* for the month of May 1961. The main line of the Mexican Railway runs from Vera Cruz to Mexico City, passing to the south of the Orizaba mountain range. There is a

Fig 14 *The* Caledon *of the Toronto, Grey & Bruce Railway, Canada. Built by the Avonside Engine Co in 1871. 3 ft 6 in gauge*

AVONSIDE ENGINE COMPANY

Fig 15 *Locomotive of the Poti & Tiflis Railway, Georgia. Built by the Avonside Engine Co in 1871. 5 ft gauge*

continuous ascent from Cordova (altitude 2,713 ft). The rise continues even more steeply up to the summit of the line at Bocca del Monte (altitude 7,923 ft). From this point the line runs along an elevated plateau to Mexico City (altitude 7,350 ft).

These Fairlie locomotives were used on the section from Cordova to Bocca del Monte, 108 miles, and in a report of 1872 it was stated that 'The Lechatelier steam brakes on the Fairlie locomotives are found to be sufficiently powerful to control the trains in descending the Orizaba incline, where the speed is limited to eight miles per hour'. A further order was executed for the Mexican Railway in 1874, and these locomotives (Works Nos 950-957) are illustrated by Fig 16. They were larger than the original series of 1871, and had cylinders of 16 in diameter by 22 in stroke; wheels 3 ft 9 in diameter; bogie wheelbase 8 ft 6 in; total wheelbase 30 ft 3 in. The boiler barrels had a length of 11 ft 4 in and a diameter of 3 ft $10\frac{3}{4}$ in at the front ring. The total heating surface was 1,816 sq ft, and the grate area 28 sq ft. Capacity of water tanks was 2,100 gallons. Like the previous three, these locomotives were woodburners and they carried railway numbers 28 to 31.

Eight 0–6–6–0 locomotives were built during 1871-2 for the Nitrate Railway. These (Works Nos 886-893) carried railway numbers 9-16, and had cylinders 15 in diameter by 22 in stroke; wheels 3 ft $7\frac{1}{2}$ in diameter; bogie wheelbase 8 ft; total wheelbase 29 ft 1 in. Total heating surface was 1,607 sq ft, grate area 24 sq ft, and working pressure 140 psi.

Total weight was 81 tons. Nos 9, 10 and 15 were still in service in 1932.

Six more were supplied in 1874 (Works Nos 944-949) (Fig 17) and one in 1880, and these received railway numbers 17-22 and 32 respectively. These locomotives were generally similar to those of 1872, but had cylinders $15\frac{1}{2}$ in diameter by 20 in stroke, and were later rebuilt by the company with new cylinders and outside Walschaerts valve gear.

Another Canadian line to use the Fairlie type was the Glasgow & Cape Breton Railway, a 3 ft-gauge system in Nova Scotia. Three of the 0–4–4–0 type were sent out in 1872 (Works Nos 907-912), and these locomotives (Fig 18) had 11 in diameter cylinders with a stroke of 19 in; wheels 3 ft 3 in diameter; bogie wheelbase 5 ft 6 in; total wheelbase 21 ft 4 in. Length of boiler barrels was 9 ft 6 in, diameter 3 ft $0\frac{1}{4}$ in; total heating surface 922 sq ft and grate area 13·25 sq ft. The capacity of the water tanks was 1,000 gallons and the coal bunkers held 1 ton 3 cwt.

Following the success of the *Little Wonder*, built at Hatcham in 1869, a second double-boiler locomotive was designed by G. P. Spooner (C. E. Spooner's son), who was engineer to the Festiniog Railway from 1872 to 1879, and this was built in 1872 at Bristol (Works Nos 929-30), being numbered 8 and named *James Spooner* after the engineer who had originally laid out the line from Portmadoc to Festiniog. The cylinders were $8\frac{1}{2}$ in diameter by 14 in stroke; coupled wheels 2 ft 8 in diameter; bogie wheelbase 4 ft 6 in, the bogies being pivoted off-

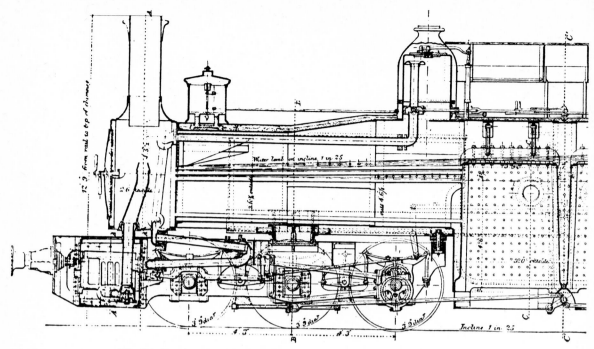

Fig 16 *Locomotive built in 1874 for the Mexican Railway by the Avonside Engine Co*

centre in india-rubber-lined bearings to counteract the weight of the cylinders; total wheelbase 18 ft 8 in. Each boiler barrel was 2 ft 7 in diameter and 7 ft 7 in long, containing 102 tubes of 1½ in diameter and 7 ft 10¼ in long, while the centre-line of the boiler barrels was 4 ft 0½ in above rail level. The inside fireboxes were 2 ft 1 in wide by 2 ft 8½ in long and 3 ft 2½ in deep; the width of the outside casing being 2 ft 7¾ in and its length 6 ft 4 in. There were two ashpans instead of one as fitted to the *Little Wonder*. Firebox heating surface amounted to 84 sq ft; the two sets of tubes contributed 629 sq ft, making a total of 713 sq ft. The grate area was 11·2 sq ft. Steam pressure was 140 psi, giving a tractive effort of 5,410 lb. Capacity of the water tanks was 720 gallons and weight in full working order, 20 tons 1 cwt.

As originally built this locomotive (Fig 19) had cast-iron stove-pipe chimneys and steam domes surmounted by Roltz safety valves; the dome covers were of polished brass. The sandboxes were mounted on the boiler barrels close behind the chimneys. On top of each sandbox was a large brass bell, and on top of the frame supporting the bell was a whistle; this rather attractive combined fitting is well illustrated in the photograph. A drawing in the Festiniog archives, dated 1888, indicates that part of the original boiler was then considered to be unsafe, and Neilson & Co supplied a new one of the parallel barrel design, in steel, for the sum of £390. This was delivered and fitted during 1889. At the same time an all-over cab and standard sandpots were fitted. This locomotive was again rebuilt in 1908 when a new boiler with tapered barrels was supplied together with cast-iron chimneys, vacuum brake, and new side tanks. It was withdrawn in 1929 and used as a source of spare parts for the other Fairlie locomotives still in use.

This locomotive was equipped with G. P. Spooner's patent regulating gear, the handles of which can be seen through the square window of the weatherboard in the photograph. It will be appreciated that a double-boiler Fairlie locomotive has two regulators, one for each bogie, situated in the domes, with separate rodding carried on bearings along the tops of the boilers and terminating in two adjacent levers. To enable these levers to be operated simultaneously, G. P. Spooner devised his patent wheel-type regulator, illustrated in Fig 20. Each rod had a lever fixed on its end, the bosses of these levers being

AVONSIDE ENGINE COMPANY

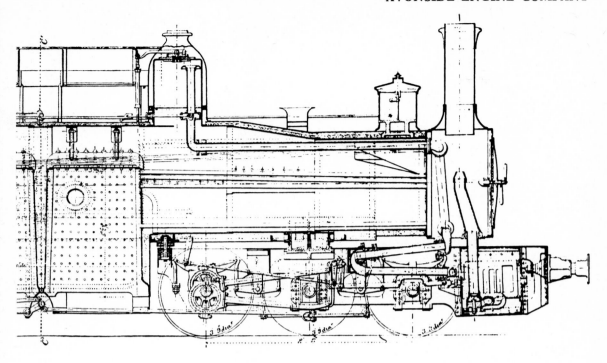

Fig 17 *Locomotive for the Nitrate Railway. Built in 1874 by the Avonside Engine Co*

THE FAIRLIE LOCOMOTIVE

very long so that bearings could be formed in them as shown. Each of the levers was fitted with a spring-loaded catch, actuated by the pivoted handle at the top, and these catches entered between the teeth of a wormwheel, the axis of which turned in the bosses of the two levers. Into this wormwheel there was also geared a worm fixed on a transverse shaft, provided at its other end with a handwheel on the driver's side of the firebox. It will be seen from the drawings that the wormwheel, being locked or prevented from rotating by the worm, served as a catch-plate for the regulating levers, and by lifting the catches by knocking the top handles into a horizontal position either of the levers could be actuated independently. Alternatively, when the top handles were put into the upper quadrant position and the catches were down or entered into the teeth of the wormwheel, the two levers could be actuated, and the two regulators thus opened or shut at the same moment by rotating the handwheel on the wormshaft. This ingenious idea does not seem to have been very successful on the Festiniog line where the rails were often wet and greasy, causing each unit of the locomotive to slip independently of the other, an occurrence that imposed a big strain on the bogie cradles. This gear therefore had a limited use, and although it was also fitted on two other Fairlie locomotives on this railway, it was removed from all three at an early date.

During the period 1873-5, an 0–4–4–0 locomotive was put into service on the narrow-gauge Matanzas Railway, Havana, in the Island of Cuba. This was constructed to the same general arrangement drawings as the *James Spooner* on the Festiniog Railway. According to V. Röll in *Enzyklopädie des Eisenbahnwesens* this locomotive was built by the Avonside Engine Company. The cylinders, bogie wheelbase, total wheelbase, and the wheel diameter were identical with the Festiniog locomotive, but the tube heating surface amounted to 536 sq ft and the firebox heating surface to 58 sq ft, so making a total of 594 sq ft. The tanks held 600 gallons, and the total weight was 21 tons. The gauge of this railway was 2 ft 6 in.

Three 0–6–6–0 locomotives for the 3 ft gauge, with cylinders $13\frac{3}{4}$ in diameter by 18 in stroke, are recorded in the official register as having been sent to Spain in 1873 (Works Nos 958-963); one of these carried the name of *Escalador de Montes* (Fig 21).

On page 11 of *Narrow Gauge Railways in America* by G. Hardy and P. Darrell, published in 1949, there appears the statement that the *Escalador de*

Fig 18 *Locomotive for the Glasgow & Cape Breton Railway, Nova Scotia. Built by the Avonside Engine Co in 1872. 3 ft gauge*

AVONSIDE ENGINE COMPANY

Fig 19 *The* James Spooner *of the Festiniog Railway. Built in 1872 by the Avonside Engine Co. 1 ft 11½ in gauge*

Montes was built for the Venezuelan Government. This book is a reprint of Fleming's original of 1877, which appeared so soon after the locomotives were built that the original author was in a good position to know where the three were sent. In view of this it seems reasonable to suppose that all three locomotives went to South America, and were perhaps exported through an agent in Spain.

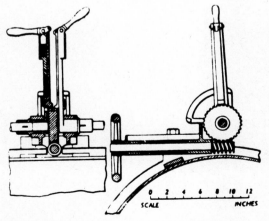

Fig 20 *G. P. Spooner's Patent Regulating Gear, as fitted to three Fairlie locomotives on the Festiniog Railway*

Two 0-4-4-0 locomotives were built for the 3 ft 6 in-gauge lines in New Zealand during 1873-4. The first, built in 1873 (Works Nos 1022-3) was named *Snake* (Fig 22), and was for the General Government, while the second, built in 1874 (Works Nos 1044-5) was named *Lady Mordaunt* and was for the Otago Provincial Council; the order for this having been placed through the Fairlie Engine & Rolling Stock Co. These locomotives were the first in New Zealand to have the Walschaerts valve gear, and had cylinders 9 in diameter by 16 in stroke; wheels of 3 ft 3 in diameter; bogie wheelbase 4 ft 6 in; total wheelbase 19 ft 6 in. The total heating surface was 766 sq ft, with a grate area of 12·5 sq ft. The water tanks held 700 gallons, and 70 cu ft of peat could be carried in the bunkers. The weight was 23 tons. The *Lady Mordaunt* was scrapped in 1896, and *Snake* was withdrawn in 1884 and dismantled in 1890.

The Nitrate Railway in Peru received three 0-6-6-0 locomotives from the Bristol works in 1873 (Works Nos 1024-1029). These had 16 in diameter cylinders with a stroke of 22 in, and 3 ft 9 in diameter wheels. The order for these was received through Bailey, Hawkins & Co as agents, and they may have been originally intended for the Iquique or Pisagua lines. At this point it may be mentioned that while records of fifteen Avonside Fairlies built

THE FAIRLIE LOCOMOTIVE

for the Nitrate Railway survived in the archives of the Administration of the Ferrocarril Salitrero in Iquique in 1932, nothing is known as to what happened to the earlier ones from the Iquique and Pisagua Railways that were absorbed into the Nitrate Railways (note the plural after 1882). The British vice-consul at Iquique made inquiries for the author and in a letter in March 1959 stated that no records of the early companies could now be found. The Nitrate Railways Company ceased to exist after 1955 when the system was taken over by the Chilean Government.

The Northern Railway of Montevideo, in Uruguay, was a local line which ran from the city of Montevideo to the village of Santa Lucia. The principal traffic consisted of meat from the abattoirs at the latter place, established there in connection with the cattle industry. This railway possessed two 0–6–6–0 locomotives named *Montevideo* and *Santa Lucia* (Works Nos 1032-1035) (Fig 23) and although the date of building is recorded in the register as 1874, the complete set of original working drawings, which are still preserved in the Peñarol Works of the State Railways, are all dated between January and April 1878. Perhaps the eight is a mistake for a three.

The cylinders were 11 in in diameter by 16 in stroke, and one unusual feature for a British-built locomotive was the use of a link motion with outside eccentrics. From the photograph it is not clear whether Stephenson or Allan gear was used. The very rare photograph reproduced here was discovered in the home of an ex-railwayman in Montevideo during 1959. One of these locomotives was reported to be out of service in 1908.

A further series of 3 ft 6 in-gauge 0–4–4–0 locomotives (Fig 24) was built in 1875 for the New Zealand Government (Works Nos 1060-1071) and intended for service in Otago, but later they were used for construction work on the North Island Main Trunk Railway. The cylinders were 11 in in diameter by 18 in stroke, with Walschaerts valve gear; wheels 3 ft 3 in diameter; weight between 34 and 37 tons. As built they had balloon stacks with a sandbox at the base of each; later they were given extended smokeboxes and handsome flanged chimneys. Numbered 177, 176, 178, 174, 173 and 172 respectively, this class was particularly unfortunate and gave constant trouble with the flexible steam pipes, and they also had a habit of fracturing their frames when working on the severely curved Wanganui section. Scrapping of these locomotives began in 1898, but the last one, No 174, survived out of use at the New Plymouth Sheds as late as 1920.

In 1875 one 0–6–6–0 locomotive was built for South Africa (Works Nos 1096-7) to the order of the Crown Agents for the Colonies, and intended for use on the 3 ft 6 in-gauge lines of the Cape

Fig 21 *The* Escalador de Montes, *built in 1873 by the Avonside Engine Co and used on the Venezuelan Government Railways. 3 ft gauge*

AVONSIDE ENGINE COMPANY

Fig 22 *The* Snake *for the General Government of New Zealand. Built by the Avonside Engine Co in 1873. 3 ft 6 in gauge*

Fig 23 *The* Santa Lucia *built by the Avonside Engine Co for the Northern Railway of Montevideo in 1874*

THE FAIRLIE LOCOMOTIVE

Government Railways. This locomotive (Fig 25), when sent out from England, was accompanied by a special engine-driver, a Mr Edmund Roberts, who had had considerable experience with Fairlie locomotives in Peru. The cylinders had a diameter of $11\frac{1}{2}$ in by 18 in stroke, with Walschaerts valve gear, and this was the first recorded use of this type of gear in South Africa. The wheels had a diameter of 3 ft 3 in on a bogie wheelbase of 7 ft 6 in, the total wheelbase being 29 ft. Total length over buffers was 45 ft 2 in. The heating surface of the tubes was 908 sq ft, and that of the fireboxes 90 sq ft, making a total of 998 sq ft. The firegrate area was 14 sq ft, and the boiler working pressure was 135 psi. The centre-line of the boilers was 6 ft 2 in from rail level and the total height to top of chimneys was 12 ft 6 in. The side tanks held 840 gallons of water and the bunkers 42 cwt of coal. The total weight in working order was 36 tons.

This locomotive was ordered at the same time as two Stephenson Patent back-to-back tank locomotives, built by Kitsons of Leeds, and a series of comparative tests were carried out on the Cape Eastern system out of East London. The Fairlie locomotive was the most economical on coal and water, and gave easier riding, but Colonial coal from Indwe and Molteno was used and this led to difficulties due to the high ash content and heavy clinker. This locomotive was scrapped about 1903. Mr John D. Tilney, locomotive superintendent, in his report of 1877 stated that for any future order for Fairlie type locomotives certain modifications were necessary, and suggested (1) that the Roscoe lubricator be fitted to the ball and socket joints; (2) that the sandboxes, which were fitted to the bogie frames, were ineffective, having too much complicated gearing—sandboxes fitted on top of the smokeboxes would utilise a simpler form of gearing; and (3) that safety chains should be fitted, or some such means of preventing the bogies from swinging round too far should the locomotive leave the rails.

When the second Fairlie locomotive for the Eastern section of the Cape Government Railways was built in 1878 all the above modifications were incorporated. This locomotive (Works Nos 1237-8), illustrated in Figs 26 and 27, had cylinders of $11\frac{1}{2}$ in diameter with a stroke of 18 in, the valves again being operated by Walschaerts gear. The wheels had a diameter of 3 ft $3\frac{1}{2}$ in on a bogie wheelbase of 7 ft 6 in, and the total wheelbase was 25 ft 2 in, the overall length being 37 ft 6 in over buffers. The bogie pivots were located 3 in to the firebox side of the centre axles, and were 17 ft apart. With the centre-line 5 ft $5\frac{1}{4}$ in above rail level, each boiler barrel had a length of 10 ft and a diameter of 2 ft $11\frac{1}{2}$ in, the total tube heating surface amounting to

Fig 24 *No 176 of the New Zealand Government Railways. Built in 1875 by the Avonside Engine Co. Class 'E'. 3 ft 6 in gauge*

36

AVONSIDE ENGINE COMPANY

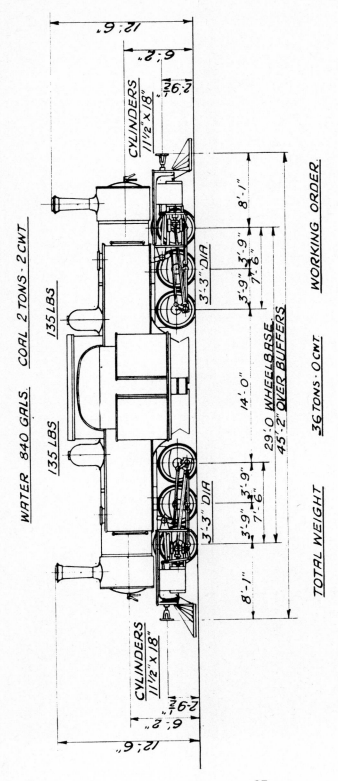

Fig 25 The first Fairlie locomotive for the Cape Government Railways, No E7, built in 1875 by the Avonside Engine Co. 3 ft 6 in gauge

THE FAIRLIE LOCOMOTIVE

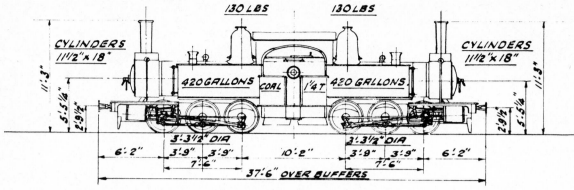

Fig 26 *The second Fairlie locomotive for the Cape Government Railways, No E33, built in 1878 by the Avonside Engine Co. 3 ft 6 in gauge*

908 sq ft. With 90 sq ft provided by the fireboxes, the total heating surface was 998 sq ft. Grate area was 14 sq ft; boiler working pressure 130 psi. The tanks held 840 gallons, and the capacity of the coke bunkers was 42 cu ft, or approximately 1¼ tons. The sandboxes were located round the bases of the chimneys, and the overall height to chimney top was 11 ft 3 in. This second Fairlie was also scrapped in 1903.

Five 0–6–6–0 locomotives (Works Nos 1115-1124) were built during the years 1875-6 for the Tamboff & Saratoff Railway, on the Russian gauge of 5 ft, and these had cylinders 15 in diameter by 20 in stroke, with 3 ft 6 in diameter wheels, and were generally similar to those built in 1871 by Sharp, Stewart & Co for the same railway. These locomotives were all sold to the Poti & Tiflis Railway in 1887.

Eighteen locomotives of the 0–6–4 single-boiler Fairlie pattern were supplied to the order of the New Zealand Government Railways, who saw in this type a design that would solve many operating problems. Of these, fifteen (Works Nos 1217-1231) were built during 1878, followed by three more (Works Nos 1232-1234) during 1879. It is understood that the Avonside Engine Company was given the contract on the recommendation of Robert Fairlie, and the price per locomotive was £1,840. These locomotives (Fig 28) were officially known as Class 'R' and had cylinders 12¼ in diameter by 16 in

Fig 27 *The second Fairlie locomotive of the Cape Government Railways, No E33, photographed at East London in 1903*

AVONSIDE ENGINE COMPANY

stroke, with the Walschaerts valve gear; coupled wheels 3 ft in diameter; coupled wheelbase 6 ft 9 in; with bogie pivot 6 in to the rear of the centre axle; distance between pivots of front and rear bogies 15 ft 8½ in; trailing bogie wheels 3 ft in diameter on a wheelbase of 4 ft 6 in. The total heating surface was 608 sq ft, and the grate area 12 sq ft. Boiler pressure was 130 psi. The water tanks held 760 gallons. Some of these locomotives were fitted with a most peculiar form of feed-water heater; in this

erected. In 1891, Nos 215, 216 and 217 were sold to the West Australian Government Railways for use on the old Eastern Main Line of that system. All three were scrapped in 1900.

With their small coupled wheels the 'R' and 'S' classes had been primarily designed to haul mixed trains at speeds of about 25 mph, but in service they proved capable of attaining speeds of up to 53 mph. These single-boiler Fairlies were very successful locomotives and some of the 'R' series remained in

Fig 28 *Single-boiler Fairlie locomotive for the New Zealand Government Railways. Built 1878-9 by the Avonside Engine Co. Class 'R'. 3 ft 6 in gauge*

system, hot ashes were caught in a large balloon-type chimney and blown down a short external pipe (near the chimney) into an internal pipe surrounded by the water in the side tanks, and were then ejected through a larger pipe which discharged into the air near the back rim of the chimney. The abrasive action of this device must have been considerable. A photograph of one of these locomotives so equipped will be found on page 9 of *The Locomotive* for 1921.

This class was followed immediately by a further series of seven almost identical locomotives (Fig 29) designated Class 'S', due to the cylinders having a diameter of 13 in instead of 12¼ in. Four appeared in 1880 (Works Nos 1279-1282) and three in 1881 (Works Nos 1283-1285) although, due to primitive workshop facilities in New Zealand at this period, it was 1886 before the last locomotive had been

Government service until the 1930s, while two that had been sold to private owners were at work as recently as 1945. One example, No R 28, is now on display at the Reefton Nursery Play Centre. Reefton is a coal-mining town on the west coast of the South Island. The last of the 'S' class was scrapped in 1926.

The first railway to be opened in Western Australia was the mineral line from Geralton to Northampton, over 300 miles from Perth, the State capital. The railway itself was thirty-three miles long and was constructed by the Western Australian Land Company. The line was taken over by the Government in 1896. To work this isolated system two Fairlie locomotives of the unusual 2-4-4-2 wheel arrangement (Fig 30) were purchased in 1879 (Works Nos 1239-1242).

These had cylinders of 10 in diameter by 18 in

THE FAIRLIE LOCOMOTIVE

stroke, with Walschaerts valve gear; leading wheels 2 ft diameter; coupled wheels 3 ft 3 in diameter; wheelbase, leading wheels to front coupled axle, 5 ft; coupled wheelbase 5 ft; total wheelbase 30 ft 6 in. The bogie pivots, located 1 ft 9 in from the front coupled axles, were large hollow spheres 12 in in diameter resting on cup-shaped sockets on the bogie cross-members, the distance between the two bogie pivots being 17 ft. The length of time that these two unusual Fairlies remained in service is not known.

The introduction of the Fairlie locomotive into India and Burma was due to the Third Afghan War of 1879, when a light metre-gauge railway was projected through the Bolan Pass, and in anticipation of this twenty-five 0-4-4-0 locomotives were ordered from the Avonside Engine Company. This line was not proceeded with, as a better route up the Harnai Valley was found and this could be worked by normal designs of tender locomotive. As, however, seventeen Fairlie locomotives had by this time been built (Works Nos 1245-1278), they were accepted by the Government of India, and shipped during 1880-81 to Bombay. Only sixteen actually arrived and were numbered in the metre-gauge stock of the Indian State Railways. It appears that one locomotive was lost at sea when the ss *Corsica* was wrecked off the coast of Portugal.

These locomotives had cylinders 11 in in diameter with a stroke of 17 in; wheels 3 ft diameter; and a weight of 36 tons 16 cwt. A Bombay Government letter of May 1881 says: 'It appears that seven Fairlie Engines with and four without, tanks have been received. One of the latter has been already erected...' (It is suggested that this refers to the first eleven arrivals.) Again, an Indian Government letter of July 1881 says: 'I am to say that these Fairlie Engines were sent out without tanks as it was found that the wheel load with tanks would be excessive. The bunkers should not, therefore, be converted into water tanks, but water should be supplied from a separate tender or tank on a wagon.' The maximum metre-gauge axle loading was 8 tons. All this suggests that only the first seven or so ever had side tanks, the others being shipped from England without them—presumably when the Indian Government realised that they would not be needed on the original line and would have to be used on ordinary metre-gauge lines. One of these seven locomotives is shown in Fig 31.

In 1881 seven were sent to Khandwa for working the Ghat section of the Holkar State Railway (later part of the Rajputana-Malwa State Railway). Only two were erected, and six-wheeled tenders were provided to carry the feed water. From the photograph (Fig 32) it will be seen that when supplied without

Fig 29 *Single-boiler type Fairlie locomotive for the New Zealand Government Railways. Built 1880-81 by the Avonside Engine Co. Class 'S'. 3 ft 6 in gauge*

AVONSIDE ENGINE COMPANY

side tanks the alterations consisted of short fuel bunkers, with holes in the cab side-sheets; sandboxes astride the boilers; and feed-supply pipes carried below the footplate from each buffer beam to the injectors. The tender had hoses at each end, so could be coupled to either end of the locomotive. The other five remained unerected. In 1885 eight were sent up from Bombay to work a temporary 1 in 18 incline on the Hirok to Kotal section of the North Western State Railway. They were joined by the two from the Holkar line, and then by the five hitherto unerected ones held by the Rajputana-Malwa Railway. All fifteen were withdrawn from this work in 1887 and put into store at Sukkur, and there offered for sale. The sixteenth locomotive had previously been sold to a contractor in Bombay. In Burma, the metre-gauge Mandalay-Kulon section of the Burma State Railways was completed in 1895, and for working this line the Indian Government sold ten of the locomotives then in store at Sukkur. Four were erected in Burma in 1896, two in 1897, three in 1898, and one in 1899. As the Mandalay-Kulon line was laid with 50 lb rail, permitting an axle load of 9 tons 4 cwt, it is presumed that the

Fig 30 *The unique 2–4–4–2 type built in 1879 by the Avonside Engine Co for the Geralton-to-Northampton line of the Western Australian Land Company. 3 ft 6 in gauge*

Fig 31 *Metre-gauge locomotive built by the Avonside Engine Co in 1880 for the Indian State Railways*

THE FAIRLIE LOCOMOTIVE

first six locomotives placed in service were those with side tanks, for it was not until October 1897 that a request was made to the Indian Government for the supply of ten tenders, and the North Western Railway of India was instructed to send the ten tenders for the Fairlie locomotives 'now at Lahore' —ie the tenders were at Lahore.

Four Fairlies remained in store at Sukkur until 1907 when they were sold to the Madras Railway for ballasting and construction work on the Nilgiri Mountain Railway. In 1908 the control of this line passed to the South Indian Railway and these locomotives were then used to operate mixed trains over the adhesion section between Coonoor (altitude 5,616 ft) and Ootacamund (altitude 7,228 ft), a distance of twelve miles. Due, however, to the advent of new motive power, these were only used for a few years, and then withdrawn about 1914.

When sold to the Madras Railway tenders were not used, and a photograph of one of these locomotives when working on the Nilgiri Mountain Railway, reproduced in *The Locomotive* for 1914, page 288, indicates that four water tanks had been formed by extending the bunker plating to a point approximately in line with the outer axle of each bogie.

Fig 32 *Metre-gauge locomotive with tender for the Indian State Railways. Built by the Avonside Engine Co in 1880-81*

CHAPTER 6

THE YORKSHIRE ENGINE CO LTD SHEFFIELD, 1872-1906

THE records of the Yorkshire Engine Company were very inaccurately kept prior to the year 1909, when a complete register of all the locomotives that had been built was drawn up. By that time, however, much data had been lost, but it has now been established that during the early 1870s Fairlie locomotives were built to stock orders for a selling agency —gauge modifications and even wheel arrangements being altered according to demand—this being the reason for the incomplete records of this type.

Five 0-6-6-0 locomotives were built for the Mexican Railway in 1872 (Works Nos 170-174), and five more in 1873 (Works Nos 190-194) (Fig 33) for the same company. These had cylinders of 16 in diameter by 22 in stroke; wheels 3 ft 6 in diameter; bogie wheelbase 8 ft; total wheelbase 29 ft 5½ in. The total heating surface was 1,688 sq ft, of which the fireboxes contributed 141 sq ft. The grate area was 26·6 sq ft. The side tanks held 2,200 gallons, but the capacity of the bunkers was only 30 cwt of coal or 180 cu ft of wood. The fireboxes were designed to burn either coal or wood, coal being obtainable at one end of the line and wood at the other.

One of these locomotives ran trials on the Grange Colliery branch of the Manchester, Sheffield & Lincolnshire Railway in February 1872 in the presence of forty engineers, including Robert Fairlie. Also present were the Duke of Sutherland, His Excellency Nicholas Novosselsky, the Mayor of Odessa, M. Landeburg, CE, of Sweden, and M. Illimoff of St Petersburg.

In 1874 one locomotive of the 0-6-6-0 pattern was supplied to the Nitrate Railway. This was numbered 23 by the company, and is thought to have been Works No 175. It had cylinders 15 in in diameter by 20 in stroke; wheels 4 ft 7 in diameter; a total heating surface of 1,468 sq ft, and a grate area of 22 sq ft.

The Hallsberg-Motala-Mjolby Railway in Sweden

Fig 33 *The* Aculcinco, *built by the Yorkshire Engine Co in 1873 for the Mexican Railway*

THE FAIRLIE LOCOMOTIVE

received two 0–4–4–0 locomotives in 1874 (Works Nos 176 and 177). These were first numbered 5 and 6, but when this private line was taken over by the State in 1879, they became Nos 285 and 286, and received the names of *Berserk* and *Nidhögg* (Fig 34).

The cylinders, which had a diameter of 13 in and a stroke of 20 in, were slightly inclined and had inside valve chests actuated by Stephenson gear. The wheels had a diameter of 3 ft 9 in, and the bogie wheelbase was 6 ft, while the total wheelbase was 25 ft 6 in. The boilers had a tube heating surface of 1,250 sq ft, which together with 119 sq ft in the two fireboxes gave a total of 1,369 sq ft. The grate area was 20·5 sq ft, and the working pressure 120 psi. These locomotives were originally intended to burn peat, and the bunkers had a capacity of 80 cu ft; the water tanks held 1,450 gallons. The total weight was 48 tons 2 cwt.

They were used mainly for goods traffic, but were not very successful machines. In 1887 the two locomotives were rebuilt by Nydqvist & Holm of Trollhaten into four small 0–4–2 side-tank locomotives, and these were finally withdrawn between 1899 and 1907.

One 0–4–4–0 locomotive was sent to Rio de Janeiro in 1873 for service on the 3 ft 7⅜ in-gauge Canta-Gallo Railway. This locomotive (Fig 35) had cylinders 12½ in diameter by 20 in stroke; wheels 3 ft 9 in diameter; bogie wheelbase 6 ft; total wheel- 24 ft 3½ in. The total heating surface amounted to 1,099 sq ft, and the grate area to 17 sq ft. Water capacity was 1,000 gallons, and the bunkers held 30 cwt.

The locomotive illustrated in Fig 36 was built in 1872 (Works No 212) for the Grande Compagnie du Luxembourg, and was the only example of the type in Belgium. It carried the name *Fenton* (after a Mr W. Fenton of Rochdale, who was president of the board of directors of the old Luxembourg Railway Company) and numbered 108 in the stock. The dimensions were as follows: cylinders 15 in diameter by 22 in stroke; wheels 3 ft 6 in diameter; total heating surface 1,580 sq ft; working pressure

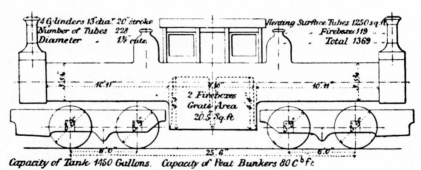

Fig 34 *Locomotive for the Hallsberg–Motala–Mjolby Railway, Sweden. Built by the Yorkshire Engine Co in 1874*

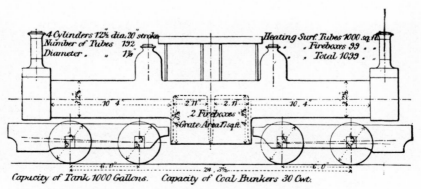

Fig 35 *Locomotive for the Canta–Gallo Railway, Brazil. Built by the Yorkshire Engine Co in 1873. 3 ft 7 5/16 in gauge*

YORKSHIRE ENGINE CO LTD

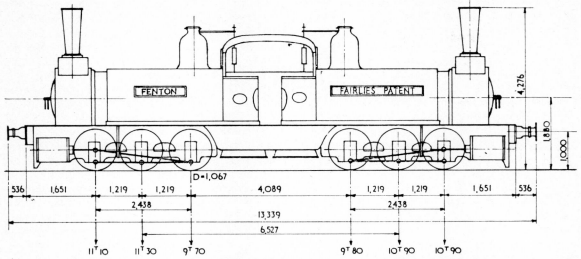

Fig 36 *The* Fenton, *built in 1873 by the Yorkshire Engine Co for the Grande Compagnie du Luxembourg. Dimensions are in millimetres*

130 psi; weight 63·7 tons. When this railway was taken over by the Belgian State Railways in 1878 this locomotive became No 969 and was used principally at Brussels, hauling local goods trains between the depots at the Alle-Verte and the Quartier Leopold. Finally it was sent to the yard at Arlon, and was scrapped in 1887.

Works Nos 219–228 were allotted to ten large 0–6–6–0 locomotives built in 1874. These had cylinders 17 in diameter by 22 in stroke, wheels 3 ft 9 in diameter, a total heating surface of 2,075·5 sq ft, a grate area of 33·5 sq ft, and a working pressure of 160 psi. The water tanks held 2,660 gallons, and the bunkers 2¼ tons. The weight empty was 55 tons 11 cwt, and 77 tons when in working order. Five were fitted with axles suitable for the 5 ft gauge, and these (Works Nos 219–223) (Fig 37) went to the Poti & Tiflis Railway in Georgia. As only four of these eventually appeared on the Transcaucasian system, one may have been lost at sea. The other five (of 4 ft 8½ in gauge) remained in stock and in *The Engineer* for 16 August 1878 the following

Fig 37 *Locomotive for the Poti & Tiflis Railway, Georgia. Built in 1874 by the Yorkshire Engine Co. 5 ft gauge*

THE FAIRLIE LOCOMOTIVE

advertisement appeared: 'For sale, five Fairlie locomotives built by the Yorkshire Engine Company, Sheffield... quite new... apply George A. Cade, 8 Old Jewry, London'. It was not until 1881, however, that a purchaser was found for them, and they were then dispatched to the Nitrate Railways in 1882.

Before being shipped to Peru, they were modified by the addition of Bissel trucks at the outer ends of the bogies (Fig 38); this alteration increased the unladen weight to 63 tons, and the weight in working order to 85 tons. These locomotives (Works Nos 224-228) became Nos 33 to 37 on the Nitrate Railways. Confirmation of this rebuilding and sale will be found in a report in *Engineering* for 7 August 1885, page 33, where it is said that 'on the Iquique Railway, Peru, probably the largest locomotives in the world are running. They weigh 85 tons in working order; have 12 coupled wheels in two groups and cylinders 17 in in diameter by 22 in stroke. The engines are provided at each end with a two-wheeled Bissel truck'. At some unknown date the railway company removed the trucks and modified the bogie frames, thus converting the locomotives back to 0-6-6-0, and their dimensions according to the Administration of the Ferrocarril Salitrero in 1932 were then: cylinders 17 in diameter with a stroke of 22 in; wheels 3 ft 10½ in diameter; bogie wheelbase 8 ft 6 in; total wheelbase 31 ft 11 in. Heating surface, tubes 1,740 sq ft, fireboxes 199 sq ft, total 1,939 sq ft. Grate area 32 sq ft. Working pressure 150 psi. From letters unearthed from the Yorkshire Engine Company's archives in 1962 it appears that in September 1881 Robert Fairlie called at the Meadow Hall Works to inspect the five locomotives then undergoing alterations, and one of them appears to have been stripped right down for detailed examination, which points to its having been put to some use between 1874 and 1881.

It has been suggested that this locomotive was the one that had been used for demonstration purposes on the East & West Junction Railway during 1876-8 and illustrated here in Fig 39. This drawing, and the photograph published in *The Locomotive* for 1911, page 251, indicates that except for modifications to the chimneys, cab and footsteps, it was indeed identical with the series, Works Nos 219-228, as built, and this theory, put forward by the late P. C. Dewhurst, may well be correct. Three more locomotives of this same design and dimensions were later supplied to the Nitrate Railways, two in 1889 (Works Nos 442 and 443) railway Nos 63 and 64, and one, railway No 73, in 1906.

During 1883 three 0-6-6-0 locomotives were built for the Mexican Railway (Works Nos 365-367) and these had cylinders 16 in in diameter by 20 in stroke, with Allan straight-link motion; wheels 3 ft 9 in diameter; bogie wheelbase 8 ft 3 in; total wheelbase 30 ft 3⅛ in. The total heating surface was 1,660 sq ft. Water capacity of the tanks was 2,350 gallons, and the coal bunkers held 1 ton 10 cwt. The photograph (Fig 40) which illustrates this class, should be compared with Fig 58, which shows the series built by Neilson & Co during the same year for the same railway. The two series were identical in outward appearance although the Neilson-built ones

Fig 38 *Locomotive built by the Yorkshire Engine Co in 1874 as 0-6-6-0 and converted by the makers in 1881 to 2-6-6-2 type for sale to the Nitrate Railways, Peru*

YORKSHIRE ENGINE CO LTD

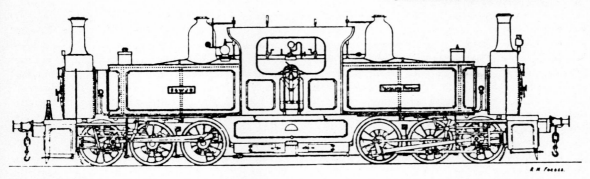

Fig 39 *Yorkshire Engine Co 0–6–6–0 of 1874 as used on the East & West Junction Railway before conversion to 2–6–6–2 type for sale to Nitrate Railways, Peru*

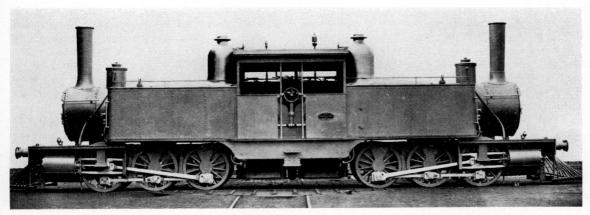

Fig 40 *Locomotive for the Mexican Railway. Built in 1883 by the Yorkshire Engine Co*

Fig 41 *Locomotive for the Anglo-Chilian Nitrate & Railway Co. Built by the Yorkshire Engine Co in 1891. 3 ft 6 in gauge*

THE FAIRLIE LOCOMOTIVE

had larger cylinders and boilers, together with a larger coal and water capacity.

The Anglo-Chilean Nitrate & Railway Company operates about 184 miles of 3 ft 6 in-gauge track connecting the two large natural nitrate plants, Maria Elena and Pedro de Valdivia, with the port of Tocopilla, and reaching an altitude of about 4,500 ft, but in contrast with the Nitrate Railways only two Fairlie-type locomotives were used. These locomotives, built in 1891 (Works Nos 446 and 447) are here illustrated in Fig 41, and had cylinders of 14 in diameter by 20 in stroke; wheels 3 ft in diameter, on a bogie wheelbase of 6 ft 10 in; total wheelbase 25 ft 10 in. Total heating surface 1,224·4 sq ft. Grate area 19·3 sq ft. With 1,100 gallons of water and 1 ton 6 cwt of coal, the weight was 52 tons 4 cwt. Numbered 8 and 9 by the company, these were withdrawn about 1929.

The Junin Railway is a 2 ft 6 in-gauge line which, like all the railways serving the nitrate area of Chile, ascends the coastal plateau. The ascent from sea-level is so steep that the railway begins at the top of the escarpment at an altitude of 2,080 ft, and is connected with the port of Junin by an aerial ropeway. From the coastal terminus at Alto de Junin to Reducto the line climbs 1,680 ft in 31·8 miles by gradients of 1 in 35 and curves of 328 ft radius, and to operate heavy trains over this section the company ordered two 0-6-6-0 locomotives of the modified-Fairlie type with two entirely separate boilers, carried on a single deep girder frame.

Built in 1905 (Works Nos 834 and 835) and named *San Antonio* and *Compania* respectively, these locomotives had the following principal dimensions: cylinders $12\frac{1}{2}$ in diameter by 16 in stroke; wheels 2 ft 6 in diameter, placed inside the frames; bogie wheelbase 6 ft; distance between bogie pivots 21 ft $10\frac{1}{2}$ in; total wheelbase 29 ft $4\frac{1}{2}$ in. The boilers had a combined tube heating surface of 930 sq ft, and the two separate fireboxes, of the Belpaire pattern, provided a further 117 sq ft, making a total of 1,047 sq ft. The grate area was 21·66 sq ft. Boiler pressure was 160 psi. The centre-line of the boilers was only 4 ft $10\frac{1}{2}$ in above rail level. The weight, with 1,500 gallons of water and 2 tons of coal, was 52 tons 2 cwt.

The difficulty arising in connection with the reversing of the outside Walschaerts valve gear, due to one end of the swing-link being attached to the carrier frame and the other to a moving bogie frame, was obviated by locating the reversing shafts at the top of the tanks. A very long swing-link was thus obtained and there was practically no slip in the die due to the movement of the bogies when rounding curves. A photograph of these locomotives appeared in *Articulated Locomotives* by L. Weiner, 1930, page 153, and in *The Locomotive* for 1907, page 8. In these two locomotives the arrangement of the steam pipes was greatly improved, and simplified: the steam being taken from the side of the boiler barrels through a copper pipe which terminated in a ball exactly under the centre of the bogie pivot, where it fitted into a socket on the end of another pipe which ran directly to the cylinders. There was thus only one ball joint, correctly placed in a position where there was a minimum of movement. The exhaust pipe was fitted with a ball at each end and worked in a corresponding socket in the smokebox bottom and the breeches-pipe between the cylinders. The cylinder drain cocks were operated by a steam cylinder, connected to a shaft to which all the cocks on one bogie were coupled.

CHAPTER 7

THE VULCAN FOUNDRY LIMITED NEWTON-LE-WILLOWS, 1872-1911

THE first two Fairlie locomotives to be built by the Vulcan Foundry were for the 3 ft 6 in-gauge Dunedin & Port Chalmers Railway in New Zealand in 1872. These were of the 0–4–4–0 wheel arrangement (Works Nos 636 and 637), being named *Rose* and *Josephine* respectively. They are illustrated by Fig 42. The cylinders were 10 in in diameter by 18 in stroke; wheels 3 ft 9 in diameter; bogie wheelbase 5 ft; total wheelbase 19 ft 9 in. Each boiler barrel had a length of 8 ft 6 in and a diameter of 3 ft 9¼ in, the total tube heating surface being 762 sq ft, and that of the fireboxes 70 sq ft, making a total heating surface of 832 sq ft. Grate area was 10·25 sq ft. The boiler pressure was 130 psi. The water tanks held 960 gallons, and the coal bunkers 17 cwt. In 1876 both locomotives were taken into the stock of the New Zealand Government Railways, and the *Rose* ceased work during the 1880s; but the *Josephine* had a much longer and more varied existence. It was sold in 1900 to the Public Works Department and used on construction work until 1907, afterwards being laid aside in a shed at Mangaweka until 1917, when it was disposed of to the Otago Iron Rolling Mills for scrap. Fortunately it was not broken up and so was overhauled and reconditioned during 1925 for the Dunedin and South Seas Exhibition, and the photograph shows this locomotive outside the Otago Early Settlers Hall where it has been on permanent display since that date. The two large balloon chimneys are not original and were only fitted in 1925.

The Patillos Railway, a 2 ft 6 in-gauge line in Peru, received two 0–4–4–0 locomotives in 1872 (Works Nos 638 and 639) (Fig 43), and these were named *Patillos* and *Lagunas* respectively. The cylinders, the total heating surface, and grate area were

Fig 42 *The* Josephine *built in 1872 by the Vulcan Foundry Co for the Dunedin & Port Chalmers Railway, New Zealand. 3 ft 6 in gauge*

THE FAIRLIE LOCOMOTIVE

the same as the two New Zealand locomotives described above, but the tanks held 900 gallons and the bunker capacity was 16 cwt. The wheels had a diameter of 3 ft 3 in. These were followed by two more for the same railway in 1873 (Works Nos 682 and 683), and according to the records they were built to the order of Lemonius & Company of Liverpool, who were presumably agents for this South American line. These locomotives had somewhat larger dimensions: the tube heating surface was 759 sq ft, and that of the fireboxes 80·8 sq ft, making a total of 839·8 sq ft. The grate area was 11·2 sq ft. The total wheelbase was 20 ft 1 in, and the weight empty 22 tons 16 cwt 2 qr.

When General Palmer carried the 3 ft-gauge track of the Denver & Rio Grande Railroad through the La Veta Pass, Colorado, at an altitude of 4,700 ft, on its way from Denver to Mexico City via Santa Fe, with grades of up to 211 ft per mile, the physical characteristics of this line were thought to be very favourable to the use of the Fairlie design of articulated locomotive. The locomotive (Fig 44) was built in 1873 (Works No 672) and is said to have been ordered by the Duke of Sutherland, who presented it to the railroad company. The *Mountaineer* was a wood-burner, weighing 29 tons 10 cwt in working order and was fitted with the Le Chatelier counter-pressure brake. It was withdrawn from service in 1883. Following this came three 0–4–4–0 locomotives for the Pimental & Chicklaya Railway, a 3 ft-gauge line in Peru. These locomotives (Fig 45) were named *Chicklaya*, *Pimental*, and *Batta* respectively (Works Nos 678-680). These three, together with the one for Colorado, were all very similar in general dimensions to those for New Zealand and the Patillos Railway.

In 1874 a 0–4–4–0 locomotive (Fig 46) named *Robert Fairlie* (Works No 734) was completed to the order of the Fairlie Engine & Rolling Stock Company for the 3 ft 6 in-gauge lines of the Norwegian State Railways. For some reason not now known, the Norwegian Government would not accept this locomotive, and it was then passed over to the Fairlie Engine & Rolling Stock Company, who sent it to Australia in 1876, carrying Fairlie builder's plates No 543 of 1876. By that time it had been altered superficially in appearance (Fig 47), although by whom is not known. It was shipped to Queensland and given a trial on the Southern & Western Railway before a decision to purchase it was made. It went into service on 25 October 1876. The name *Governor Cairns* would appear to have been attached to the locomotive when it went into service, as it was not the practice of the Queensland Government at that time to name locomotives. In 1882, the running number 41 was allotted to it, and it retained this number for the remainder of its working life. A full report and comment on the transfer of this locomotive to Queensland appeared in *Engineering* for 2 March 1877, page 171, where the leading dimensions were stated to be as follows: cylinders 11 in diameter by 18 in stroke; wheels 3 ft 3 in diameter; bogie wheelbase 5 ft; total wheelbase 20 ft 8 in; length over buffers 32 ft 10 in; length of fire-

Fig 43 *The* Patillos *of the Patillos Railway, Peru. Built in 1872 by the Vulcan Foundry Co*

THE VULCAN FOUNDRY LIMITED

Fig 44 *Denver & Rio Grande Railroad, Colorado. The* Mountaineer *of 3 ft gauge, built in 1873 by the Vulcan Foundry Co*

boxes 6 ft 6 in; outside width of fireboxes 3 ft 2½ in; length of boiler barrels 8 ft; inside diameter 2 ft 11½ in; tube heating surface 856 sq ft; fireboxes 85 sq ft; total heating surface 941 cu ft; grate area 15 sq ft; tank capacity 750 gallons; coal carried 1 ton; total weight empty 26 tons 12 cwt, weight in working order 34 tons 12 cwt. In an official rolling stock list of 1879, the tank capacity is given as 960 gallons, and the capacity of the coal bunker as 1 ton 5 cwt. This locomotive was used, as far as can be ascertained, on the range climb to Toowoomba on the Southern & Western Railway. During the latter years of its life it was used mainly for hauling coal trains from the Ipswich area to the South Brisbane coaling wharf near the Wooloongabba railway yards. In 1902 it was sold to P. A. Barbat & Company, then a general engineering firm in Ipswich, who are said to have cut it up and made it into two stationary winding engines. One boiler portion is said to have been sold to a sawmill at Mooloolaba, on the

Fig 45 *The* Pimental *built by the Vulcan Foundry Co in 1873 for the Pimental & Chicklaya Railway, Peru. 3 ft gauge*

THE FAIRLIE LOCOMOTIVE

Fig 46 *The* Robert Fairlie *built by the Vulcan Foundry Co in 1874 (Works No 734) to the order of the Fairlie Engine & Rolling Stock Co for the Norwegian State Railways. 3 ft 6 in gauge*

near North Coast, but the accuracy of this cannot be verified. The other section of the locomotive may have gone to a local colliery. One of these winding engines, some years later, was at Blackheath Colliery on the Ipswich coalfield, and the frame of a driving bogie was still lying in the bush close to that colliery as recently as 1963.

The North Wales Narrow Gauge Railway, of 1 ft 11½ in gauge, from Dinas Junction (on the Caernarvon to Afon Wen section of the LNW) to South Snowdon, with a branch to Bryngwyn, was opened for goods traffic in May 1877, and the first two locomotives were single-boiler Fairlies with one power-bogie, built by the Vulcan Foundry, the first of the 0–6–4 tank wheel arrangement in the British Isles. They were built in 1875 (Works Nos 738 and 739).

The author has copies of the original drawings, and they are signed 'C. E. Spooner per G. PERCIVAL SPOONER, 12th., January, 1874'. There is also an

Fig 47 *The Vulcan Foundry Co locomotive, Works No 734 of 1874, as rebuilt and renamed* Governor Cairns *and sent to Queensland, Australia, in 1876. 3 ft 6 in gauge*

THE VULCAN FOUNDRY LIMITED

Fig 48 *The* Snowdon Ranger *of the North Wales Narrow-Gauge Railway. Built by the Vulcan Foundry Co in 1875. 1 ft 11½ in gauge*

oval stamp on the drawings which reads 'NORTH WALES NARROW GAUGE RAILWAY, ENGINEER'S OFFICE, PORTMADOC'. Mr C. E. Spooner held the position of engineer to both the Festiniog Railway and the NWNGR. These two locomotives (Fig 48) bore the names of *Snowdon Ranger* and *Moel Tryfan* respectively, and were very similar in design to the four-coupled locomotive introduced by G. P. Spooner on the Festiniog Railway about a year later (see Fig 50). The cylinders were 8½ in diameter by 14 in stroke, with Stephenson valve gear; coupled wheels 2 ft 6 in diameter, on a wheelbase of 6 ft; rear bogie wheels 1 ft 7 in diameter, on a wheelbase of 3 ft 6 in; total wheelbase 14 ft 11½ in. The boiler barrel was 2 ft 0½ in in diameter and 8 ft long between tubeplates, containing 104 tubes of 1½ in diameter. The total heating surface was 366 sq ft, grate area 5·9 sq ft, and working pressure 160 psi. The power-bogie was pivoted upon a cast-iron saddle riveted to the underside of the boiler. Steam was

Fig 49 *The* Rio Douro *built in 1875 by the Vulcan Foundry Co for the Porto a Pavoa de Varzim Railway, Portugal. 2 ft 11½ in gauge*

THE FAIRLIE LOCOMOTIVE

delivered to the valve chest through a pendulum pipe, pivoted immediately below the smokebox, and in order to avoid excessive movement in the top flexible joint this pipe was continued down into a chamber behind and below the valve chest, where it terminated in a second flexible union. The exhaust system was similar, but owing to the cylinders being in advance of the smokebox, the pipe was arranged to slope back into the blastpipe. The original drawings of these two locomotives are now in a very worn condition, and it is regrettable that it has not been possible to illustrate the steam and exhaust pipes in detail; the same applies to the drawings of the Festiniog locomotive *Taliesin* of 1876. The weight in working order was 9 tons 10 cwt on the coupled wheels, and 4 tons 10 cwt on the trailing bogie. The water capacity was 350 gallons. As built, no continuous brake appears to have been fitted, but both locomotives underwent some form of rebuilding at the works of Davies & Metcalfe Limited of Romiley, near Stockport; the *Snowdon Ranger* being dealt with in 1902, and the *Moel Tryfan* in 1903, and the Westinghouse brake, with which both locomotives were equipped in their later years, may have been fitted at that time. The Westinghouse compressor was installed in the cab and the air reservoirs beneath the footplate on either side of the cab floor.

About 1917, both locomotives were found to be in a very worn condition, and as the frames of *Snowdon Ranger* were rather better than those of *Moel Tryfan* they were put under the sister locomotive, and *Snowdon Ranger*'s boiler, cab and tanks, and *Moel Tryfan*'s frames cut up for scrap. The reconstructed hybrid locomotive continued in service until 1936, when it was taken into the Boston Lodge Works of the Festiniog Railway for boiler repairs, but this work was never carried out and the locomotive was dismantled some years later, and its remains were to be found scattered around the yard until 1955, when most of these parts were broken up, though the trailing bogie survived in 1958.

Two 0–4–4–0 locomotives (Fig 49) were exported during 1875 to Portugal for service on the 2 ft 11½ in-gauge Porto a Pavoa de Varzim Railway. Built to the order of the Fairlie Engine & Rolling Stock Company, they were named *Rio Douro* and *Rio Aye* (Works Nos 740 and 741) and had cylinders 10 in in diameter by 18 in stroke, with wheels of 3 ft 3 in diameter. In Volume Five of Heusinger Von Waldeck's *Specielle Eisenbahntechnik*, 1883, there is a long description of this line and the working conditions of these locomotives. In recent times and until withdrawn only a few years ago, these locomotives were employed on the seven-mile branch from the junction at Senhora da Horo to the beach at Matosinhos. They were apparently particularly useful for banking the densely overcrowded holiday trains on feast days.

In 1876, a single-boiler 0–4–4 type locomotive, with only one power-bogie, was supplied to the Festiniog Railway, Portmadoc, having been designed by G. P. Spooner. The original drawings, copies of which are in the author's possession, are dated

Fig 50 *The* Taliesin *for the Festiniog Railway, built by the Vulcan Foundry Co in 1876. 1 ft 11½ in gauge*

THE VULCAN FOUNDRY LIMITED

Fig 51 *Modified Fairlie type with two separate boilers, built in 1901 by the Vulcan Foundry Limited for the metre-gauge lines of the Burma State Railways*

14/10/75 and bear the signature of the designer. The power-bogie and running gear details were standardised with the bogies of the *James Spooner* (the double-boiler Fairlie built by Avonside in 1872, Fig 19), while the general dimensions, parts and nearly all the layout was identical with the two 0-6-4 single-boiler Fairlies on the North Wales Narrow Gauge Railway (see Fig 48).

This locomotive (Fig 50) was numbered 9 and named *Taliesin* (Works No 791); later, in 1883, it was renumbered 7. The cylinders were 9 in diameter by 14 in stroke; coupled wheels 2 ft 8 in diameter, on a wheelbase of 4 ft 6 in; trailing bogie wheels 1 ft 7 in, on a wheelbase of 3 ft 6 in; centre to centre of bogie pivots 9 ft 9 in; total wheelbase 13 ft 11 in. The boiler barrel was 2 ft 7 in in diameter and 7 ft 7 in long, containing 94 tubes of 1⅝ in diameter and 7 ft 10½ in long. The inside firebox had a width of 2 ft 1¾ in by 3 ft long, and was 3 ft 8 in deep at the back increasing to 4 ft 5 in at the front; the width of the outside casing being 2 ft 7¾ in by 3 ft 6 in long. A heating surface of 29·5 sq ft was provided by the firebox, the tubes contributed 313 sq ft, making a total of 342·5 sq ft. Grate area was 6·3 sq ft. The working pressure was 150 psi, and the tractive effort 3,029 lb. With a tank capacity of 380 gallons and a bunker space for 1 ton 5 cwt of coal, the weight in full working order was 17 tons. This locomotive had a flexible steam pipe and a rocking exhaust pipe, Stephenson valve gear, Salter safety valves on the dome, and the regulator was of the Stroudley pattern located in the dome. In the course of a working life of fifty-four years this locomotive was completely renewed. At the first rebuilding in 1891, new all-steel bogie wheels were fitted, together with a new smokebox and chimney. In 1894 new frames and cylinders were made, and finally in 1900 it was rebuilt with a new steel boiler, smokebox, larger tanks holding 430 gallons, an all-over cab with side windows and tool-box behind, sand pots, and vacuum brakes. The new boiler came from the Vulcan Foundry. It was a successful locomotive, but the size of the coupled wheels, which had a tendency to slip, made it unsuitable for heavy traffic and so it was used almost exclusively on light passenger trains. The boiler was condemned in 1930, and the locomotive withdrawn from service, being dismantled in 1932.

After completion of the *Taliesin* for the Festiniog Railway in 1876, no further Fairlie locomotives were built by the Vulcan Foundry during the next twenty-five years until 1901, when the firm introduced a modified form with two separate boilers carried on a single deep girder frame. One advantage of this design is that the variation of the water level on grades is reduced, although the overall length of the locomotive is increased by an amount equal to the distance between the two fireboxes. Seven of these locomotives (Fig 51) were constructed for the metre-gauge lines of the Burma Railways, of which five (Works Nos 1773-1777) were completed in 1901, and two (Works Nos 2200 and 2201) in 1906. The principal dimensions were: cylinders 14 in diameter by 20 in stroke; wheels 3 ft 3 in diameter; bogie wheelbase 7 ft 7 in; total wheelbase 35 ft 7½ in. Each boiler had a barrel with a length of 8 ft 5 in, and was 3 ft 6½ in in diameter, containing 152 tubes of 1⅝ in diameter. The combined heating surface was 1,132 sq ft in the tubes, with 138 sq ft in the two fireboxes; to this was added 128 sq ft in the Drummond patent water-tubes fitted to each firebox, thus giving a total heating surface of 1,398 sq ft. Grate area was 13 sq ft in each firebox. The water capacity was only 500 gallons, while the bunkers held 2½ tons of coal. The weight in working order amounted to 60½ tons.

THE FAIRLIE LOCOMOTIVE

These locomotives were employed on the Zibingyi ghat, and due to the small water capacity of the side tanks, used tenders for some years after their construction, these being at first taken from the Avonside Fairlies of 1880-81, but later they had other tenders of their own. The valve gear fitted to these locomotives was a modified form of Joy radial gear, in which the anchor link was attached to a small return crank in line with the main crank. The fitting of externally arranged orthodox Joy valve gear to outside-cylindered locomotives was rare at any time, but the most numerous application of this modified form was on several classes of tank locomotives built during the period 1885 to 1896 by five British makers for Japan. It was also used in Russia on a few classes of locomotive built between 1890 and 1896.

During 1911 the Mexican Railway received three 0-6-6-0 locomotives (Works Nos 2586, 2587 and 2588) designed to haul 300-ton trains up the 1 in 25 grades of this system. At that date, these large and imposing machines (Fig 52) were probably the heaviest built in Britain for the standard gauge, the total weight being 138 tons, or 23 tons per axle. The cylinders had a diameter of 19 in and a stroke of the unusual length of 25 in; wheels 4 ft in diameter; bogie wheelbase 9 ft 3 in; total wheelbase 35 ft 6 in. The tube heating surface amounted to 2,679 sq ft, and that of the two fireboxes to 245 sq ft, bringing the total heating surface to the high figure of 2,924 sq ft. The fireboxes had rocking grates and were designed for either coal or oil firing; the combined grate area was 47·7 sq ft. A single steam dome was provided on the back ring of one boiler barrel and this was equipped with four Ashton pop safety valves set to a working pressure of 185 psi. These locomotives, Railway Nos 183 to 185, were the last of the type ordered by the Mexican Railway, and it is of considerable interest therefore to note that in December 1923, just prior to the electrification of the Orizabo-Esperanza section, there were eighteen Fairlie locomotives still in service, of four different classes; all had been converted to oil firing.

Fig 52 *Locomotive No 184 of the Mexican Railway. Built by the Vulcan Foundry Limited in 1911*

CHAPTER 8

R. & W. HAWTHORN & COMPANY NEWCASTLE UPON TYNE, 1874-90

THIS Newcastle firm constructed four Fairlie locomotives of the 0-4-4-0 design during 1874 for the Nassjo-Oscarshamn Railway in Sweden. These locomotives (Fig 53) (Works Nos 1636-1639) were named *Berga*, *Malilla*, *Morlunda* and *Marianneland*, the railway numbers being 6 to 9. The cylinders had a diameter of 10 in and a stroke of 18 in; wheels were 3 ft 6 in in diameter, bogie wheelbase 5 ft, total wheelbase 19 ft 7 in; overall length 29 ft 6 in. The boiler barrels had a length of 8 ft 6 in and a diameter of 2 ft $9\frac{1}{2}$ in, the combined tube heating surface amounting to 743 sq ft, which together with 68 sq ft in the two fireboxes made a total of 811 sq ft of heating surface. The grate area was 10 sq ft. With 850 gallons of water and 14 cwt of coal, the weight in working order was 27 tons 2 cwt. Nos 6 and 7 were broken up in 1910, and Nos 8 and 9 sold to a railway contractor in 1899.

In 1876 two 0-6-6-0 locomotives (Works Nos 1689 and 1690) were delivered to the Bolivar Railway, a 2 ft-gauge line in Venezuela connecting Barquisimeto with Tucacas on the Caribbean. These had cylinders 9 in diameter by 14 in stroke; wheels 2 ft 6 in diameter, and an overall length of 33 ft 3 in. They were numbered 7 and 8 by the railway. Some years later, in 1890, a similar locomotive (Works No 2162) was supplied to the 2 ft-gauge industrial railway of the Quelbrada Copper Company in Venezuela. This had cylinders 9 in diameter by 14 in stroke; wheels 2 ft 6 in diameter;

Fig 53 *The* Malilla *built by R. and W. Hawthorn in 1874 for the Nassjo–Oscarshamn Railway, Sweden*

THE FAIRLIE LOCOMOTIVE

bogie wheelbase 6 ft; total wheelbase 22 ft 8 in; overall length 32 ft 2 in. The weight was 23 tons. It was later acquired by the Bolivar Railway and numbered 15.

The Norwegian State Railways ordered one 0-4-4-0 locomotive in 1875 (Works No 1697), and this was named *Robert Fairlie* and delivered in 1877. Built to a gauge of 3 ft 6 in it had cylinders 10 in in diameter by 18 in stroke, and the wheels had a diameter of 3 ft 3 in. The overall length was 31 ft. It was seen, partly dismantled, at Trondhjem in 1895.

The next locomotive (Works No 1699), built in 1877, was of the 0-4-4 single-boiler type with only one power-bogie, supplied to the order of the Fairlie Engine & Rolling Stock Company, and presents a mystery that has hitherto escaped notice by locomotive historians. It was constructed to exactly the same drawings as the single-boiler Fairlie that appeared on the Swindon, Marlborough & Andover Railway during 1881, and which has been referred to in many books and journals since 1900 as having been built by the Avonside Engine Co (Works No 1244); in the Avonside records at the Hunslet Engine Co, Leeds, this number is blank. Recent research has brought to light additional information that points definitely to there having been only the Hawthorn locomotive. When new in 1877 it was tried, for demonstration purposes, on the East & West Junction Railway (later the Stratford-upon-Avon & Midland Junction), but after a brief period working passenger trains between Stratford and Blisworth, it was sent to the Paris Exhibition of 1878, carrying the name *Robert Fairlie*. While at the Exhibition it was described in *Engineering* for 28 June 1878, page 511, as having been built by the Avonside Engine Co, but on the other hand, *The Engineer* for 1878, although giving a very full description, including an illustration, of this same locomotive at the Exhibition, makes no reference as to where it was built.

After the Exhibition it was put up for sale, and was available for inspection at the Machinery Register Depot, Newport, Monmouthshire. The following advertisement appeared in April 1881: 'NEW: single-boiler, double-bogie Fairlie locomotive engine, cylinders 16 in diameter, 22 in stroke; this engine took a prize at the Paris Exhibition, and is splendidly got up. £2,000'. In September 1881 it was offered on loan to the Swindon, Marlborough & Andover Railway who, the following March, offered to purchase it for £1,000. The offer was accepted in April 1882, and the locomotive became No 4 on the railway. From Fig 54 it will be seen that the steam bogie in front had outside cylinders, which were horizontal and 16 in in diameter by 22 in stroke, the valve gear being of the Walschaerts type. The travel of the valves was $4\frac{1}{2}$ in, the lap $1\frac{1}{8}$ in and the lead $\frac{3}{16}$ in. The coupled wheels had a diameter of 5 ft 6 in on a wheelbase of 6 ft 6 in, while the trailing bogie had wheels of 4 ft diameter on a wheelbase of 6 ft; the distance between the

Fig 54 *Swindon, Marlborough & Andover Railway No 4. The single-boiler Fairlie built by R. and W. Hawthorn in 1877*

R. & W. HAWTHORN & COMPANY

Fig 55 *State Railways of Saxony No 18, built in 1880 by R. and W. Hawthorn. 2 ft 5½ in gauge*

two bogie centres being 15 ft 4 in. The total wheelbase was 21 ft 7 in and the length over buffers 33 ft 2 in. The boiler, of Lowmoor iron, was made in three rings, the middle one being the largest with a diameter of 4 ft 2⅜ in; the plates were $\frac{9}{16}$ in thick. The raised firebox casing was 5 ft 1 in long, and the inside box was 4 ft 10 in long by 3 ft 8 in wide, with a depth of 4 ft 5 in; it was stayed to the outer shell by girder roof stays. There were 181 tubes of 2 in diameter, giving a heating surface of 1,022 sq ft, and the firebox provided 72 sq ft, making a total heating surface of 1,094 sq ft. The grate area was 15 sq ft. The centre-line of the boiler, which had a chimney with a polished copper cap, was 7 ft 0¼ in above rail level. A pair of Naylor patent safety valves was mounted on the firebox. Brakes were fitted to the wheels of the trailing bogie as well as to those of the power-bogie; maximum braking of this kind was very unusual at that period. The large side tanks and the tank under the coal bunker together held 1,200 gallons, while the bunker capacity was 2 tons. The total weight in working order amounted to 44 tons. This locomotive was never a success and its coal consumption was reported to have been as high as 50 lb per mile. That eminent engineer historian, the late E. L. Ahrons (1866-1926), from personal observations at the time, expressed the opinion that this excessive figure was due in all probability to inaccurate setting of the valves. The Walschaerts valve gear, at that period, was a complete novelty to running-shed fitters in Britain, although many English and Scottish locomotive building firms had long constructed machines fitted with this gear for overseas railways. From about 1883 onwards this locomotive was only steamed in emergencies, then in 1888 it was put to work for a short period on the local exchange traffic between Swindon Town and Swindon Transfer (GWR) goods yards. It was finally scrapped in 1892, the boiler being fitted up to supply steam to the company's workshops at Cirencester.

The last Fairlie locomotives to be built at the Forth Banks Works were for the State Railways of Saxony, which in 1885 purchased two of the 0–4–4–0 type (Works Nos 2012 and 2013) for working on 2 ft 5½ in-gauge lines having 1 in 33 gradients and curves of 50 metres radius, between Hainsberg and Kipsdorf, and between Alterburg and Heidenau. These locomotives (Fig 55) had cylinders 8¼ in in diameter by 14 in stroke; wheels 2 ft 8 in diameter; bogie wheelbase 4 ft 6 in; total wheelbase 18 ft 8 in; overall length 27 ft 10 in. The total heating surface was 692·6 sq ft, and the grate area 15·5 sq ft. The weight empty was 22 tons. An early photograph indicates that when new they had chimneys with large conical spark arresters. These two locomotives, which were considered to be rather expensive to operate, were satisfactory for a time, but they were ultimately replaced by locomotives of the Meyer design.

CHAPTER 9

WIENER-NEUSTADTER LOKOMOTIV-FABRIK VORMALS G. SIGL, 1879

In addition to the locomotives built by the Avonside Engine Co in 1871-2, and by the Yorkshire Engine Co in 1874, the Poti & Tiflis Railway placed five more Fairlie locomotives in service during 1879. These were built in Austria at the old SIGL works, the title of which was changed before the construction of these locomotives to Wiener-Neustadter Lokomotivfabrik vormals G. Sigl.

These locomotives (Fig 56) (Works Nos 2439-2443), carried railway Nos 109-113, and had cylinders 381 mm in diameter by 506 mm stroke; wheels 1·080 m in diameter; tube heating surface 141·4 sq m; firebox 15·6 sq m; total heating surface 159 sq m; working pressure 10·5 kilograms per sq cm; grate area 2·6 sq m; weight empty 58·6 tonnes; in working order 71·8 tonnes. Water tank capacity was 10

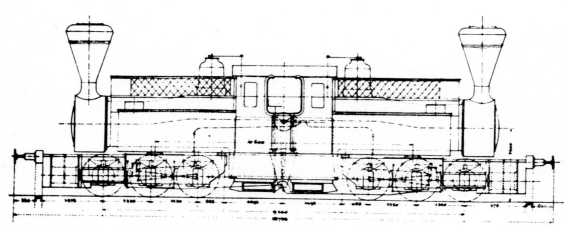

Fig 56 *Locomotive for the Poti & Tiflis Railway, Georgia. Built by Wiener-Neustadt in 1879. 5 ft gauge*

WIENER-NEUSTADTER LOKOMOTIVFABRIK

cu m, and there was storage for 8 cu m of wood fuel. The valve gear was Allan straight-link, and the tractive effort of this class was 10,270 kilograms. The drawing is from a catalogue of the Wiener-Neustadter company preserved in the Vienna Railway Museum. A photograph of No 111 of this series is reproduced on page 58 of Rakov's book on *Locomotives of the Railways of the Soviet Union*.

All records of the Wiener-Neustadt locomotive works were destroyed by fire, near St Stephen's Cathedral in Vienna, in the last days of World War II.

CHAPTER 10

FESTINIOG RAILWAY BOSTON LODGE WORKS, 1879-85

For the unusual operating conditions existing on the narrow-gauge Festiniog Railway in North Wales, the Fairlie principle had fully justified its use since 1869, and the last locomotives built for this line were two of the 0–4–4–0 double-boiler design, constructed in the company's own works at Boston Lodge, situated at the south end of the Traeth Mawr embankment and about a mile from Portmadoc. Exactly what parts were manufactured in these workshops has always been a matter of conjecture; certainly the foundry would have been capable of producing the necessary castings, and the machine shop was remarkably well equipped for such a small railway. When the works were taken over in 1915 as a National Shell Factory, the main items of heavy equipment were a planing machine, slotting machine, three lathes, a shaping machine, drilling machine, boring machine, and a half-ton steam hammer. It is, however, known that the boilers were completely fabricated in the works, although there does not seem to be any record as to whether facilities ever existed for rolling the boiler plates, and whether the necessary flanging blocks were available for shaping the firebox plates.

The first of the locomotives, No 10, *Merddin Emrys*, ran its trial trip on 21 July 1879, and the second, named *Livingstone-Thompson* and numbered 11, appeared in 1885.

The cylinders were 9 in diameter by 14 in stroke, placed 3 ft $2\frac{1}{2}$ in between centres; coupled wheels 2 ft $9\frac{1}{4}$ in diameter; bogie wheelbase 4 ft 8 in; the bogies being pivoted off-centre in india-rubber-lined bearings and fitted with a cast-iron box at the firebox end, filled with lead to counterbalance the weight of the cylinders. The bogie frames had a length of 10 ft $2\frac{1}{4}$ in and were 1 ft $4\frac{1}{2}$ in deep and $1\frac{1}{8}$ in thick. Centres of bogie pivots 15 ft. Total wheelbase 20 ft.

These locomotives had taper boilers when built, the barrels being in three rings $\frac{3}{8}$ in thick. Each barrel was 2 ft $6\frac{1}{4}$ in in diameter at the smokebox end and 8 ft $1\frac{3}{8}$ in long, containing 124 tubes $1\frac{1}{2}$ in in diameter and 8 ft $4\frac{1}{4}$ in long. The inside fireboxes were each 2 ft $1\frac{1}{4}$ in wide at the bottom by 2 ft $10\frac{1}{2}$ in long at bottom. Height to crown above centre-line of boiler 7 in. The width of the outside casing was 2 ft $7\frac{1}{4}$ in, and it was 6 ft 8 in long. The heating surface of the fireboxes was 69·9 sq ft, tubes 817·2 sq ft, making a total of 887·1 sq ft. Grate area 12·1 sq ft. Working pressure 160 psi. Tractive effort was 6,059 lb. Each of the original tanks measured 7 ft 9 in long by 1 ft 5 in wide and 2 ft $7\frac{1}{2}$ in deep, the four tanks holding a total of 667 gallons, but these were later extended forward to the line of the smokeboxes. Coal capacity was 1 ton 5 cwt, and it is recorded that the consumption was 10 cwt for the round journey of 26 miles 36 chains, or about $45\frac{1}{2}$ lb per mile. Width over footplates 6 ft 6 in. Weight empty 18 tons 6 cwt, and in working order 24 tons.

As originally built these locomotives (Figs 57 and 57a) had Spooner type wheel-operated regulators, Salter safety valves, and cabs. No 10 had stovepipe chimneys and (a feature of unusual interest for the period) copper water-tubes supporting the brick arches, and columnar pattern sandboxes. No 11 had plain fireboxes and small rectangular shaped sandboxes in front of the tanks. Both locomotives had steel wheels and tyres formed as one-piece castings.

No 10 was rebuilt in 1896 with new frames, new steel wagon-top boilers, new smokeboxes, new capped chimneys and new cab. In 1921 it received another new boiler of similar type constructed by the Vulcan Foundry Limited, while in 1934 the boiler and firebox received extensive repairs at the Avonside Engine Company.

No 11 was later renumbered 3 (probably during 1886), and in 1908 was rebuilt with a new steel wagon-top boiler supplied by the Vulcan Foundry Limited, and new smokeboxes and chimneys were

FESTINIOG RAILWAY

Fig 57 *The Merddin Emrys of the Festiniog Railway. Built at the Boston Lodge Works of the company in 1879. 1 ft 11½ in gauge*

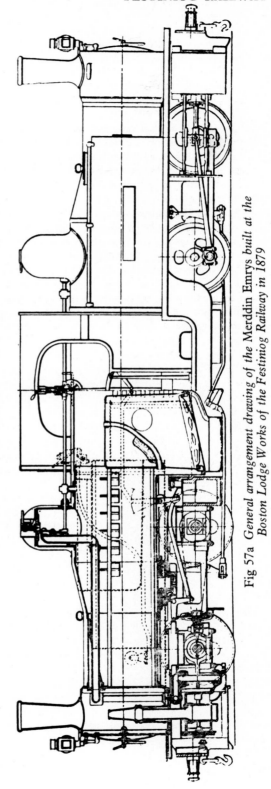

Fig 57a *General arrangement drawing of the Merddin Emrys built at the Boston Lodge Works of the Festiniog Railway in 1879*

THE FAIRLIE LOCOMOTIVE

fitted at the same time. In 1931 this boiler received extensive repairs by the Avonside Engine Company. At the first rebuilding, the new boilers of both locomotives had barrels 8 ft $1\frac{3}{4}$ in long and 2 ft $7\frac{3}{4}$ in in diameter, and other modifications were the fitting of the vacuum brake and the removal of the wheel-operated regulators. The name of No 11 was changed to *Taliesin* in 1930. It is now named *Earl of Merioneth*. The Festiniog Railway Company suspended operations on 2 August 1946, and the line lay moribund, overgrown with vegetation, until the formation of the Festiniog Railway Preservation Society Limited, a company formed to ensure the preservation of the railway and to render financial assistance to the railway company. A train service over a limited distance commenced during the summer of 1955, and at the time of writing (1969) the line is again open to Dduallt. Early in 1956 the Vulcan Foundry Limited was asked to survey these two Fairlie locomotives with a view to making one of them serviceable; both were found to be worthy of repair, but the *Taliesin* was already in the erecting shop, where it had been since 1940 awaiting tubeplate repair; it was decided to recondition this one on site with spare parts made in the Vulcan shops. The bogies were already partly repaired but some other components were removed and sent to Newton-le-Willows for renewal. The boiler barrels, which had to be re-tubed and re-stayed, were repaired on site, but new smokeboxes were made at the Vulcan shops. The locomotive was then re-assembled and steamed in September 1956, and was placed in regular service during the 1957 season. All this work was carried out by Vulcan fitters, as the railway had no shop personnel working at that time.

No 10, which had been in the locomotive shed since 1946, was later similarly rebuilt and returned to traffic in 1961, but this rebuilding had in fact taken two years; on this locomotive the cab was not replaced and the only protection for the enginemen is the two weatherboards. The chimneys are of a new pattern, with a pronounced taper. During 1968, two new boilers equipped with superheaters were ordered for both these locomotives from the Hunslet Engine Company of Leeds.

CHAPTER 11

NEILSON & COMPANY, HYDE PARK WORKS, GLASGOW, 1883-1901

BETWEEN 1883 and 1901, this famous Scottish firm supplied fifteen Fairlie locomotives to the Mexican Railway; all were of the 0-6-6-0 design. In 1883 three were built (Works Nos 2874-2876) (Fig 58) and these carried railway Nos 156-158. They were similar in external detail and appearance to the three supplied to this railway by the Yorkshire Engine Company in the same year, although these Neilson ones had larger cylinders and boilers, together with a larger coal and water capacity (see Fig 40).

The cylinders were of 16 in diameter by 22 in stroke, with Allan valve gear; wheels 3 ft 9 in diameter; bogie wheelbase 8 ft 3 in; centres of bogie pivots 22 ft; total wheelbase 30 ft 3 in. The diameter of each boiler barrel was 3 ft $10\frac{3}{4}$ in at the smokebox end, enlarged to 4 ft $6\frac{1}{16}$ in for the ring next to the firebox shell. The outer shell for the two inner fireboxes was 9 ft long and 4 ft $1\frac{1}{8}$ in wide. Each internal firebox measured 4 ft $1\frac{5}{8}$ in long by 3 ft 6 in wide. There were 144 tubes of $1\frac{7}{8}$ in outside diameter in each boiler barrel. The total length of the complete twin boiler was 31 ft 8 in between smokebox tube plates. The heating surface of the two fireboxes was 166 sq ft, and that of the tubes 1,647 sq ft. Total heating surface 1,813 sq ft. The grate area was 29·4 sq ft. Working pressure 165 psi.

The two bogies were connected by a carrier frame which took all the buffing and traction stresses. Each bogie pivot consisted of a short gunmetal cylinder, 20 in in diameter, fixed to a cross-member on the carrier frame and fitting over a shallow cast-iron socket fastened to the bogie. To check any tendency to pitching movements, each bogie was steadied at the inner end by a helical spring, which formed an elastic connection between the inner end of the bogie frame and a transverse plate on the carrier frame close to the firebox. Between the boiler and cylinders the steam was conveyed through ball-and-socket jointed pipes. The bottom of the smokebox was of cast iron, formed with junction pieces for both the steam and exhaust pipes. The steam pipe, of copper $3\frac{3}{4}$ in in diameter, was flange-jointed to a

Fig 58 *Locomotive No 157 of the Mexican Railway. Built in 1883 by Neilson & Co*

THE FAIRLIE LOCOMOTIVE

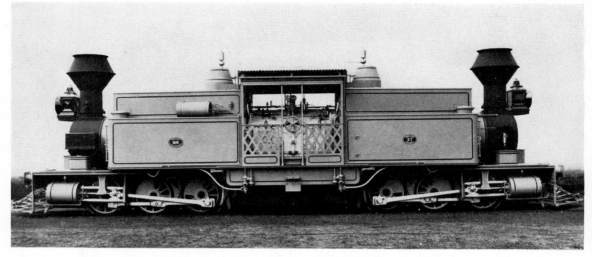

Fig 59 *Locomotive No 27 (later re-numbered 166) of the Mexican Railway. Built by Neilson & Co in 1894*

knee-pipe cast in the bottom of the smokebox, this knee-pipe being connected by a 3¾ in diameter wrought-iron pipe to a knuckle-chamber, fixed to the bogie frame, whence the steam passed forward by branch pipes to the two cylinders. The connecting pipe was made with a ball-and-socket joint at each end. The exhaust steam was conveyed to the blast pipe in the smokebox through a 5 in diameter wrought-iron pipe with ball-and-socket joints.

These were large locomotives and their weight in working order, with 2,850 gallons of water and 3¾ tons of coal, was 92¼ tons. A contemporary report claimed that these locomotives could haul a train weight of 3,600 tons on a level track.

During 1889 the first of a new class of twelve locomotives appeared, and these were delivered over a period of twelve years in five series as follows: three in 1889 (Works Nos 3895-3897); three in 1891 (Works Nos 4319-4321); three in 1894 (Works Nos 4674-4676); one in 1898 (Works No 5365); and two in 1901 (Works Nos 5761 and 5762). These locomotives (Fig 59) had cylinders 16 in diameter by 22 in stroke, with wheels 3 in smaller in diameter than the previous class. The bogie wheelbase was 8 ft 3 in and the total wheelbase 32 ft 5 in. With a tube heating surface of 1,532 sq ft, and with 180 sq ft in the fireboxes, the total amounted to 1,712 sq ft. The grate area was 33 sq ft.

Although this class had the same water and coal capacity, and presented such a massive appearance, they were 5 cwt lighter than the 1883 series, the weight being 92 tons.

The whole class was numbered 159-170 by the Mexican Railway. The photograph illustrates the nine locomotives of 1889-94, but the later series were not fitted with the large spark-arresting chimneys; they were built new with extended smokeboxes.

CHAPTER 12

THE KOLOMENSKY WORKS RUSSIA, 1884

The Kolomensky Works, situated at Kolomna, seventy-two miles south-east of Moscow, was founded in 1862. The first locomotive was constructed there in 1869 and at the end of steam locomotive production in 1956, 10,450 had been built. These works also produced coaches, bridges, river steamers and stationary steam engines.

Seventeen large 0-6-6-0 Fairlie locomotives were built there for use on the Russian State Railways, and this series is illustrated in Fig 60. In Rakov's book the only date that appears in the text when describing these locomotives is 1884, and the example shown here is known to have been built in that year. The cylinders were 15 in in diameter by 20 in stroke, with inside valve gear; wheels 3 ft 6 in in diameter; total heating surface 1,682 sq ft; grate area 23 sq ft; working pressure 142 psi; weight in working orders 90·2 tonnes.

With these seventeen Russian-built locomotives, the total number of 5 ft gauge Fairlies that eventually worked most of the traffic over the Transcaucasian system was forty-five, made up as follows:

Sharp, Stewart & Co	1871	
	ex Tamboff & Saratoff Rly	10
Avonside Engine Co	1871	
	ex Poti & Tiflis Rly	4
Yorkshire Engine Co	1874	
	ex Poti & Tiflis Rly	4
Avonside Engine Co	1875–6	
	ex Tamboff & Saratoff Rly	5
Sigl	1879	
	ex Poti & Tiflis Rly	5
Kolomensky Works	1884	
	Built for Russian State Rlys	17

Fig 60 *Locomotive for Russian State Railways. Built at the Kolomensky Works in 1884. 5 ft gauge*

THE FAIRLIE LOCOMOTIVE

The principal depot where these locomotives were stationed was Baku on the Caspian Sea. The products of the oilfields at Sabunki passed by an eight-mile pipeline to the refineries at Black Town, Baku, and one of the duties of the Fairlie locomotives was the operation of trains of forty oil-tank cars from these refineries up a steeply graded line to the main distribution sidings, from whence the trains ran $561\frac{1}{2}$ miles to Batoum (Batum), a port on the Black Sea.

The reason for the concentration of these Fairlies on the Transcaucasian Railway was the inclines of the Souram Pass, in the foothills of the Lesser Caucasus, where the grades ranged from 1 in 125 to 1 in $22\frac{1}{4}$ with curves of 200 ft radius.

From about 1903, however, these articulated locomotives were gradually replaced by eight-coupled locomotives of normal build as these were found to be more economical in working.

All these locomotives were built as wood-burners, but with the development of the Russian oil industry in the area in which they worked, they were all eventually converted to oil-burners; the large fuel tanks being mounted on top of the boiler barrels, with the transverse air reservoirs on top of the fuel tanks.

Of these forty-five locomotives, forty-three survived to become Nos 9800-9842 in 1912. In 1924 to 1926 they were transferred to the Rioni-Tvkibuli line and its branches, and although most of them were withdrawn over the next few years, a number survived in use until about 1936.

CHAPTER 13

SÄCHSISCHE MASCHINENFABRIK VORMALS RICHARD HARTMAN, CHEMNITZ, GERMANY, 1902

In 1902, the Saxon Engine Works Limited, of Chemnitz, supplied three unique Fairlie locomotives of the 0-4-4-0 type to the Saxon State Railways for working on metre-gauge branch lines connected with industrial plants. These routes had curves as sharp as 30 metres radius.

These three engines were the only compound Fairlie locomotives in the world, and had high-pressure cylinders 11 in in diameter by $14\frac{15}{16}$ in stroke driving the coupled wheels of one bogie, and low-pressure cylinders $16\frac{15}{16}$ in diameter by $14\frac{15}{16}$ in stroke driving the coupled wheels of the other bogie, the cylinder position being unusual in that all four were located at the inner or firebox end of the bogie frames. The coupled wheels, placed outside the frames, were 2 ft $5\frac{15}{16}$ in in diameter, and each bogie had a wheelbase of 3 ft $7\frac{15}{16}$ in, the total wheelbase being 24 ft $11\frac{1}{4}$ in. The boiler carried a working pressure of 199 psi, and the grate area of the single firebox was 20 sq ft. The heating surface of the firebox amounted to 86·5 sq ft, and that of the two sets of tubes to 876·5 sq ft, making a total of 963 sq ft. The capacity of the tanks was 705 gallons, and the bunker held $1\frac{1}{2}$ tons. In full working order the weight was 40 tons 2 cwt.

Since the lines ran for considerable distances along the public highways, provision was made for the driver to be always at the front by placing the

Fig 61 *Locomotive No 251 of the Saxon State Railways. Built by the Saxon Engine Works in 1902, for the metre-gauge*

THE FAIRLIE LOCOMOTIVE

Fig 62 *Former Saxon State Railways locomotive built in 1902 by the Saxon Engine Works, as working on the Deutsche Reichsbahn and renumbered 99.162*

steam and brake controls at either end. In addition to the central cab, and to enable the crew to pass safely from one end of the locomotive to the other, a large canopy with side panels extended the full length of the locomotive (Fig 61); the motion, of the Walschaerts pattern, was also covered in by side sheets. The canopies were removed at a later date (Fig 62), when it was decided that the driver should occupy the central cab.

It has been said that as with the earlier Hawthorn locomotives for the 2 ft 5½ in-gauge lines of the same railway administration, the boilers gave trouble and this in addition to the inconvenient driving position in public streets, led to their being superseded by Meyer type locomotives. This cannot have been strictly true, for these three compound locomotives were renumbered 99.161 to 99.163 by the Deutsche Reichsbahn in the 1920s, and one having been sold to Greece, the other two were still to be found working around the Reichenbach/Vogt area in Saxony in the early 1960s. The most recent photographs of these locomotives known to the writer are in two magnificent German publications: *Unvergessene Dampflokomotiven* by K. E. Maedel, 1966, and in *Liebe alte Bimmelbahn* by K. E. Maedel, 1967, where illustrations appear showing these Fairlies hauling passenger trains through the streets of Reichenbach in May 1957.

CHAPTER 14

NORTH BRITISH LOCOMOTIVE CO LTD, HYDE PARK WORKS GLASGOW, 1903-08

This company, which was formed in 1903 by the amalgamation of Sharp, Stewart & Co (Atlas Works), Neilson, Reid & Co (Hyde Park Works), and Dübs & Co (Queen's Park Works), commenced the delivery in that year of a series of ten 0-6-6-0 locomotives to the Mexican Railway. This class (Fig 63) came from the Hyde Park Works and consisted of four built during 1903 (Works Nos 16028-16031) and two built in 1904 (Works Nos 16541 and 16542) followed by the final four some years later in 1908 (Works Nos 18210, 18211 and Works Nos 18313, 18314). The railway numbers of the whole series were 171-180.

The cylinders had a diameter of 16 in with a stroke of 22 in; wheels 3 ft 6 in in diameter; bogie wheelbase 8 ft 3 in; total wheelbase 32 ft 5⅛ in. The total heating surface was 1,712 sq ft, and the grate area 33 sq ft. Working pressure 160 psi. With 2,850 gallons of water and 300 cu ft of fuel, the total weight was 99 tons.

This class was followed during 1908 by two 0-6-6-0 locomotives of much greater power and these are illustrated by Fig 64. The cylinders had a diameter of 17 in and the stroke (like that of the Vulcan series of 1911) was of the unusual length of 25 in. With wheels of 4 ft 6 in diameter the bogie wheelbase was increased to 9 ft 3 in, while the total wheelbase was 35 ft 6 in. With a tube heating surface of 2,142 sq ft, and 234 sq ft in the two fireboxes, the total was 2,376 sq ft. The grate area amounted to 43·5 sq ft, and the boiler working pressure was increased to 200 psi. As in the Yorkshire Engine Co locomotives of 1905 for the Junin Railway, the reversing shafts were located at the tops of the tanks and connected to the radius rods of the Walschaerts gear by long swing links with universal joints at either end. The large water tanks held 3,500 gallons, and the capacity of the fuel bunkers was 315 cu ft.

These fine locomotives had a weight in working

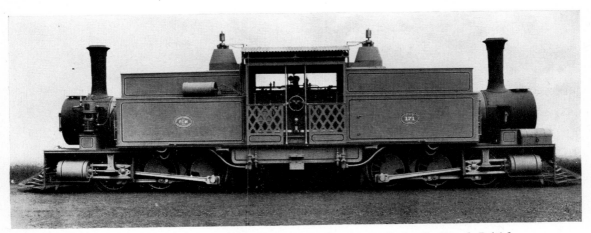

Fig 63 *Locomotive No 171 of the Mexican Railway. Built by the North British Locomotive Co Ltd in 1903*

THE FAIRLIE LOCOMOTIVE

order of 120 tons 2 cwt, and the tractive effort at 75 per cent of the boiler pressure was 45,166 lb. The Works Nos were 18315 and 18316, and they carried railway numbers 181 and 182.

Fig 64 *Locomotive No 181 of the Mexican Railway. Built by the North British Locomotive Co Ltd in 1908*

CHAPTER 15

THE HUNSLET ENGINE CO LTD, LEEDS, 1908

It will be remembered that the North Wales Narrow Gauge Railway commenced operations in 1877 with two single-boiler Fairlie locomotives. These were followed by two locomotives of orthodox design; but many years later, in 1908, a third locomotive of the single-boiler Fairlie type was constructed for this railway by the Hunslet Engine Co of Leeds. Although of the same 0-6-4 wheel arrangement as the two Vulcan Foundry locomotives, this Hunslet product (Works No 979) differed considerably in outward appearance, the boiler having a raised round-top firebox, the side tanks stopped short over the driving axle, while the rear bogie had outside frames. This locomotive is illustrated in Fig 65, and was later named *Gowrie*.

The diameter of the cylinders was $9\frac{1}{2}$ in with a piston stroke of 14 in; the coupled wheels had a diameter of 2 ft $4\frac{1}{2}$ in, on a wheelbase of 5 ft 6 in, the steam bogie being pivoted on a saddle riveted to the boiler barrel. The trailing bogie had wheels of 1 ft 10 in diameter, and the total wheelbase was 14 ft. The boiler, 2 ft 5 in in diameter and 8 ft 10 in long, contained 65 brass tubes of $1\frac{5}{8}$ in diameter outside, providing 252 sq ft of heating surface; the copper firebox had a heating surface of 30 sq ft, making a total of 282 sq ft. The grate area was 5 sq ft. The working pressure was 160 psi, with steam distribution by Walschaerts valve gear. The tanks held 400 gallons of water, and the bunker capacity was 1 ton 2 cwt of coal. The weight in working order amounted to 18 tons 10 cwt, of which 11 tons 6 cwt rested on the coupled wheels.

Fig 65 *The Gowrie of the North Wales Narrow Gauge Railway. Built in 1908 by the Hunslet Engine Co. 1 ft $11\frac{1}{2}$ in gauge*

THE FAIRLIE LOCOMOTIVE

This locomotive was not a successful design, and was reported to have been constantly short of steam when in service; there was also some trouble with the steam pipes. It is not generally appreciated how lengthy the steam pipes could be on even a small Fairlie locomotive, and in this instance they were between 8 ft and 9 ft long, causing quite an appreciable drop in pressure.

After only a few years service, this locomotive was sold in 1915 to the British Government, and at the end of World War I it turned up in Wake's Yard at Darlington in 1919; we hear of it again when Messrs Ridley & Young of Darlington offered it for sale in March 1923 for £600. Later in the same year it was used on an aerodrome contract at Marske-by-the-Sea, Yorkshire, and was last heard of in 1928 when offered for sale on completion of this contract.

The North Wales Narrow Gauge Railway, which had been extended through to Beddgelert and on to Portmadoc in 1923, and renamed the Welsh Highland Railway, was closed in June 1937. In 1941 the surviving equipment and rails were requisitioned by the Ministry of Supply and dismantled as scrap.

CHAPTER 16

MISCELLANEA

IN addition to the locomotives actually built and recorded above, the names of other railways which were said to have received Fairlies have been mentioned at times by Robert Fairlie and others, but there is no proof that such locomotives existed; presumably the details that have been published refer to specifications prepared in response to inquiries which did not materialize into definite orders.

Over a period of many years and from a variety of sources, a tradition has persisted that Cross & Co of St Helen's supplied several 0-4-4-0 locomotives to Venezuela. In *Engineering* for 12 January 1866, and in *The Engineer* for 20 April 1866, there appears a note to the effect that Cross & Co were constructing several locomotives somewhat smaller than the *Progress* for the narrow-gauge Venezuela Central Railway. In the obituary notice of Robert Fairlie in *Engineering* for 11 April 1885, these same elusive 0-4-4-0 locomotives are again mentioned, but are there said to have been disposed of to other railways. No trace of such locomotives has yet been found on any other railway.

A design for a remarkable double-power-bogie single-boiler articulated tender locomotive was prepared less than a year after the *Progress* appeared: this was illustrated in *The Engineer* for 13 July 1866 and would have been of the 0-6-4-0 type, with the cylinders at the leading end of each bogie.

In a paper read before the Society of Arts in March 1868, Fairlie described and illustrated a proposed design for a large double-boiler 0-6-6-0 locomotive that would have had cylinders 18 in in diameter by 24 in stroke; wheels 4 ft in diameter; a total heating surface of 2,550 sq ft, and a tractive effort of 33,500 lb. An engraving of this proposed locomotive appeared in the *Practical Mechanic's Journal* at the time, and was there said to have been designed for the Mexican Railway, but was probably rejected by the company because of the great weight of such a locomotive.

The extreme flexibility of the Fairlie type would seem to render it unsuitable for prolonged high speeds, yet in spite of this, two such designs were certainly prepared for locomotives intended for express passenger service.

One such specification the late A. R. Bennett discovered when compiling the classic work on early industrial locomotives, *The Chronicles of Boulton's Siding*, was for a 0-4-4-0 with 5 ft 9 in wheels and with cylinders attached to framing of the 'Crewe' or 'Allen-Buddicum' pattern. This drawing is reproduced in *The Locomotive* for 1925, page 107, and as the measurements are given in both English and metric, it is thought that the design was got out for a French railway. The drawing shows a single firebox, and so may be dated not later than 1870.

Four locomotives of the 2-4-4-2 type, with inside cylinders, and coupled wheels as large as 5 ft 6 in diameter, are said to have been ordered by the Great Russian Railway (Grande Societe of the St Petersburg & Moscow Railway) from Sharp, Stewart & Co in 1870. The late E. L. Ahrons, on page 316 of his *British Steam Railway Locomotive, 1825-1925* said that these locomotives were placed in service during 1871-2, and there have been several other indefinite references to them; however, they are not mentioned in any of Sharp, Stewart's records, and there is no room for them in the list of works numbers. They are shown in Robert Fairlie's own book, *Battle of the Gauges: Railways or No Railways*, published in 1872, but the illustration depicting this inside cylinder type is a shaded drawing, not an engraving, and cannot be accepted as evidence that such a locomotive was ever built. It is significant that in a recent very authoritative book *Locomotives of the Railways of the Soviet Union* by V. A. Rakov, published in Moscow in 1955, all the other known designs of 5 ft-gauge Fairlies in Russia are tabulated, but no mention is made of these apocryphal inside-cylinder locomotives.

In Fairlie's list in *Engineering* for 10 November 1871, giving particulars of locomotives built or on order, one 0-6-6-0 is shown as for 'Swiss Railway Co', with cylinders 16 in diameter by 22 in stroke;

75

THE FAIRLIE LOCOMOTIVE

wheels 3 ft 6 in diameter; a total heating surface of 1,740 sq ft, and a tank capacity of 2,000 gallons. This locomotive is untraceable. The Credit Valley in Canada is said, by Robert Fairlie in his article published in *Technische Mitteilungen* in 1876, to have had double-boiler locomotives in use at that date, while there is a reference in *Engineering* for 21 August 1874 to the effect that due to the fine performance of the Fairlie locomotive on the Toronto, Grey & Bruce Railway the president of the Credit Valley Railway had decided to use the type for the heavy traffic on this line. The Railway and Locomotive Historical Society of America has not been able to find any trace of such locomotives.

This same issue of *Engineering* for 21 August 1874, also contains diagrams of 0–6–6–0 locomotives for South America. Two refer to designs for the Iquique Railway and the dimensions given are as follows:

Cylinders	$15\frac{1}{2}$ in × 20 in	15 in × 20 in
Wheels	3 ft 6 in	3 ft 6 in
Bogie wheelbase	8 ft	7 ft 8 in
Total wheelbase	28 ft $11\frac{1}{2}$ in	27 ft 11 in
Total heating surface	1,725 sq ft	1,600 sq ft
Grate area	26 sq ft	
Water capacity	2,100 gallons	2,000 gallons
Coal	2 tons	

The third diagram is for an even larger design for the Pisagua Railway, with these dimensions:

Cylinders	17 in × 22 in
Wheels	3 ft 9 in
Bogie wheelbase	8 ft 6 in
Total wheelbase	31 ft $10\frac{3}{4}$ in
Total heating surface	2,059 sq ft
Grate area	33·82 sq ft
Water capacity	2,600 gallons
Coal	2 tons 10 cwt

In the above-mentioned article by Robert Fairlie, where he enumerates forty-three railways as using his patent locomotive in 1876, three railways in Peru are named; these were the Chimbote-Huaraz, the Lima-Oroya, and the Tarapaca Junction. In both Volumes III and V of Heusinger von Waldeck's *Specielle Eisenbahntechnik*, written in 1883, there is a list giving dimensions of Fairlie locomotives on twenty-four railways, and these three Peruvian lines are included. Here the Chimbote-Huaraz (a 3 ft-gauge railway) is credited with three 0–6–6–0 locomotives weighing 46 tons, while the Lima-Oroya and the Tarapaca Junction railways are given as possessing one 0–6–6–0 each, both with a weight of 64 tons. None of the South American locomotives can be accounted for in any maker's records.

CHAPTER 17

COMMENTARY

A NUMBER of constructional details of the Fairlie design are mentioned in the text in connection with individual locomotives, but some further comments are necessary to round off and conclude the story.

With the completion of the *Tarapaca* at Hatcham in 1870 the design had almost reached maturity, and after 1871 there was virtually no further change until the introduction of two separate boilers during 1901-06.

Boilers and Frames

The two boiler barrels were united by a common outer firebox casing to form one complete structure, the outer casing enclosing two inner fireboxes, separated by a water space. When working, each firebox generated steam independently according to the draught and efficiency of firing. A few boilers were made with parallel barrels, but in general the tapered wagon-top construction was used, permitting the use of raised fireboxes and a higher water level, this latter feature being of some importance on steep gradients, and with the central position of the fireboxes the variation in depth of water on the crowns of the fireboxes at the tube plates was quite small on such gradients.

All Fairlies after 1869 had ample domes on the back ring of each boiler barrel. The lower sides of the firebox outer casing were attached to the frames and each boiler barrel was cradled on the frame over each bogie centre. Every Fairlie had a short frame having a bogie pivot at each end, and this served to link together the two power-bogies; the weight of the boiler, water tanks, bunkers and cab was not carried by this frame, but directly by the pivots. As each bogie had draw- and buffing-gear at its outer end, the boiler was relieved of all strains due to operating forces.

During 1901-06 a modified form of Fairlie was introduced by both the Vulcan Foundry and the Yorkshire Engine Co, using two separate boilers with Belpaire fireboxes. Advantages claimed for this type of construction were: an unrestricted footplate; space for additional water tanks under the footplate; reduced variation of water level on gradients; less danger to the driver and fireman should the locomotive overturn. On the other hand, the disadvantage of this separate-boiler form of construction was that the overall length of the locomotive was increased by an amount equal to the distance between the two fireboxes, and this necessitated a greater distance between the centres of articulation. In the original design of boiler with an outer firebox casing joining the two barrels, a very rigid structure was formed between the two bogie pivots, but due to the loss of this rigidity with two separate boiler units, a deep girder frame was necessary to carry the two boilers. The fireboxes were supported on the frame, and the front ends of the boilers rested on strong steel castings which were fixed at the extreme ends of the frame; the undersides of these castings formed the pivots of the bogies. The bogies of all classes of Fairlie locomotive had some form of swing link at the inner end, near the firebox, connecting frame to bogie. Its purpose was to prevent pitching of the bogie about its centre; a limited degree of flexibility was allowed on the link by the use of rubber or spring buffing.

Valve Gears

In the early locomotives for the Neath & Brecon, the Anglesey Central, and the Southern & Western of Queensland, Stephenson gear was used, while on the *Little Wonder* of 1869 Gooch gear was tried. Allan's straight-link motion first appeared on the *Tarapaca* in 1870 for the Iquique Railway. Little information is available as to the gears used by the Avonside Engine Co, Sharp, Stewart & Co, or the Yorkshire Engine Co on the early series of Fairlies constructed up to about 1874, but in the latter year locomotives built for the Iquique Railway had Allan gear. Valve gears of the link type were nearly always placed inside the frames, with the steam chests having the valves on vertical faces. Notable exceptions to this were the Queensland locomotives, where the

THE FAIRLIE LOCOMOTIVE

motion was between the wheels and frame, and the two 0–6–6–0 locomotives built for the Northern Railway of Montevideo, where the two eccentrics were mounted externally on a return-crank. With the introduction of the Walschaerts gear soon after 1873, the steam chests could be placed on top of the cylinders, giving greater facility for oiling and attention to port faces.

The Mexican Railway, however, retained the inside position for the valve gear until 1908, and this may have been due to the long gradients that locomotives on this line had to coast down without steam on. With vertical port faces the valves could fall away and allow air to pass freely from one end of the cylinder to the other, thus reducing the tendency to draw grit from the smokebox. Finally, on the twin-boiler locomotives built in 1901-06 by the Vulcan Foundry for Burma, a modified design of outside Joy radial valve gear was used. The reversing of the valve motion, on bogies capable of considerable lateral movement, from one central position in the cab, created a problem that was surmounted in several ways. The locating of the reversing shafts across the tops of the boiler barrels with long swing links and universal joints to the radius rods, has already been mentioned in connection with the Walschaerts gear, but in the last Fairlies built by the Vulcan Foundry in 1911 for the Mexican Railway, a horizontal shaft ran fore and aft of the cab and then at each end a short shaft, connected by bevel gears, led down via a universal joint at the lower end, to a gearbox on a transverse shaft on each bogie frame. This transverse shaft, carrying the lifting arms, was operated by a worm and wormwheel in the gearbox. How the necessary flexibility in the reversing gear linkage was arranged on inside valve gears is not clear from the few drawings available.

Crossheads

Except for the use of single-bar crossheads on the Vulcan Foundry locomotives for Burma, all Fairlies had the normal 'alligator' pattern of crosshead with one bar above and one below. It should, however, be mentioned that a photograph of Nitrate Railways No 10, published in 1932 and illustrating this locomotive as then at work, with oil firing, also clearly shows that it then had single slide bars. Whether all the locomotives of this Avonside series, railway numbers 9-16, were so fitted when originally built in 1871-2 is not known. The Nitrate Railways carried out a considerable amount of rebuilding at the Iquique works, as witness the conversion of the Yorkshire Engine Co 2–6–6–2 locomotives to 0–6–6–0, and the fitting of new cylinders and outside Walschaerts valve gear to the 1874 series of Avonside 0–6–6–0, railway numbers 17-22.

Steam and Exhaust Piping

A degree of flexibility was required in the case of both the steam and exhaust pipes leading to and from the cylinders. Considerable detail has been given in the text regarding these fittings as applied to some seven classes between 1865 and 1905, and this will be sufficient to show the gradual improvement made, and it is therefore unnecessary to pursue this particular subject further.

Cabs

These were located over the fireboxes and firing was done through twin doors on one side. Fuel bunkers were provided fore and aft of the cab on the fireman's side, and fuel was sometimes piled on top of the tanks. When wood was used as fuel, large racks were provided extending almost to the smokeboxes. This is particularly well illustrated in photographs of the Sharp, Stewart locomotives for the Imperial Livny Railway, and the Avonside example for the Toronto, Grey & Bruce Railway. Firing was always a difficult operation due to the restricted space on the footplate, and with wood it must have been well nigh impossible to handle when in motion due to the difficulty of collecting it from the racks, and it is probable that firing was done as much as possible during stops. For the nitrate lines in South America, commencing with the *Tarapaca*, a unique cab was adopted for some years; this took the form of two separate narrow roofs on each side of the firebox casing, supported on pillars from the footplates, and having ornate lattice-work fenders. Several of the illustrations portray this form of cab, which was only suitable for use in arid climates. The other forms of cab were of two kinds: one of sheet iron with a very large opening above the fenders, and the other was a timber structure, similar to contemporary American practice. This latter form of cab had three openings on each side; the centre one formed the entrance, and the others were fixed and had windows.

Performance and Riding Qualities

The double-boiler Fairlie locomotive was particularly adaptable for use on railways where loads had to be hauled from sea level to a plateau through mountainous country with sharp curves and unusually heavy gradients. On the Nitrate Railways, the Mexican Railway, and the Transcaucasian railway system, the Fairlie type gave invaluable service for over fifty years, working traffic over arduous terrain under conditions that could not have been successfully dealt with by contemporary locomotives of orthodox type.

The double-bogie wheel arrangement gave the same steadiness as a bogie passenger coach with the ability to traverse very sharp curves, and at the same time provide maximum adhesion.

All Fairlie locomotives had wheels of small diameter, and on the six-coupled bogies the centre pair of wheels was flangeless.

The Nitrate Railways served the mines working the large deposits of sodium nitrate, and these were situated on a plateau between the coastal range of mountains and the Andes running parallel to them. From sea-level at Iquique to the summit, a distance of 32 km, the average grade was 1 in 33 with curves of 270 ft radius or more. On reaching the flat pampa, the Fairlies exchanged the train with normal type tank locomotives for the remainder of the journey to the mines. When descending the inclines, the trains were checked by brakesmen riding the wagons, and with the old type link-and-pin couplers it will be appreciated that the locomotives must have been subjected to very heavy buffing surges. Two remarkable demonstrations of the stability of the design occurred on the Mexican Railway. In the first instance, recorded in *Cassiers Magazine* for August 1896, a train was descending an incline when the brakes failed and the locomotive and wagons got out of control. Some of the wagons were derailed and wrecked, the locomotive then broke loose and dashed off on a stretch of six or seven miles down the slope, over reverse curves of 350 to 400 ft radius. It reached the bottom safely and was then pulled up. The speed is believed to have been as high as 70 mph. One of the curves traversed is more than a semi-circle, and is of about 400 ft radius, while the gradient there is even worse than 1 in 25. The locomotive was one of the Neilson series with 3 ft 6 in diameter wheels, and having regard to the terrific speed at which the motion must have been working and the swaying of the whole machine during the run, the fact that nothing gave way forms one of the finest testimonials that the Fairlie system could ever receive. The second accidental test also took place on the Mexican Railway about 1900, when a Fairlie locomotive which had been left standing at the top of an incline, with the brakes improperly secured, started down the grade and ran for no less than 48 km (thirty miles) and finally came to rest without derailing.

As many as three Fairlies could be used on a train and no difficulties arose on up grades, as on the Souram Pass of 1 in $22\frac{1}{4}$ on the Poti & Tiflis Railway in Georgia, but on downhill running precautions had to be taken where the trailing bogie of one locomotive was coupled to the leading bogie of the following one. With steam shut off, the individual oscillations of the two adjacent bogies could interact with sufficient violence to derail one of the bogies. This could be avoided by supplying just sufficient steam to the cylinders of the leading locomotive to maintain tension of the drawbar, and to regulate the degree by a light application of the handbrake to maintain the minimum pull necessary to check oscillation. The same damping effect on the tendency to oscillate could be attained by coupling a wagon between the two adjacent bogies.

* * *

The single-boiler Fairlie, with only one power-bogie, was more suitable for passenger work at moderate speeds, and as an example, reference may be made to the work of the 0–6–4 'R' class on the New Zealand Government Railways. These were supplied in 1878-9 for use on mixed trains over routes with less formidable gradients (1 in 50), but with sharp curves where a flexible wheelbase was essential. Although primarily designed to haul mixed trains at a maximum speed of 25 mph, locomotives of Class 'R' were occasionally used on express trains. On 11 July 1879, a demonstration train was run from Wellington to Summit and back to test their capabilities, the train consisting of eight wagons, a passenger car and a brake van. Instructions were given to the driver that the run was to be made as quickly as possible, but to keep within the limits of safety, and on the outward journey speeds of 45 mph were reached between Petone and Upper Hutt. On the return journey, with the load reduced to the passenger car and the brake van, the thirty-five miles from Summit to Wellington were run in sixty minutes; the last twenty miles from Upper Hutt took only thirty-two-and-a-half minutes. A maximum of 53 mph was maintained for two miles. Considering that the driving wheels had a diameter of 3 ft $0\frac{1}{2}$ in this was indeed a remarkable performance. The journey in both directions was timed by the late Charles Rous-Martin, an early railway enthusiast still remembered in Britain by the older generation. The normal loading of a Class 'R' Fairlie over the 1 in 50 gradients was nineteen wagons.

CHAPTER 18

THE MASON-FAIRLIE LOCOMOTIVE, 1871-1914

IN America the construction of locomotives on the Fairlie principle was taken up by William Mason of the Mason Machine Works, Taunton, Massachusetts, who devoted a great amount of time, energy and thought to the development of these locomotives.

The first was a large 0–6–6–0 double-boiler locomotive (Fig 66) named the *Janus*, from the god of that name, facing backwards and forwards. The works number was 438, and the date of construction is generally considered to be December 1871, although an engraving of this locomotive appeared in *Engineering* for 16 July 1869. It had cylinders 15 in in diameter by 22 in stroke; wheels 3 ft 6 in in diameter; and a total heating surface of 1,696 sq ft; but it was lack of fuel and water capacity that prevented the locomotive from being a success. Originally built for the Central Pacific Railroad in California it was first tried on the Boston & Albany Railroad and then eventually sold to the Lehigh Valley Railroad where it was used on pusher service on the Wilkes-Barre Summit; its final fate was to be made into two 0–6–0 shunting locomotives.

The *Janus* was an unsuccessful experiment and thereafter only Fairlie locomotives of the single-boiler design were built, modified so as to incorporate bar-framing and other American features, and these were designated 'Mason-Fairlies'.

The *Onward*, Works No 461 (Fig 67), was completed in 1871, and sold on 1 July 1872 to the American Fork Railroad in Utah, and was the first Mason-Fairlie to be built. This was for a gauge of 3 ft, and it had cylinders 10 in in diameter by 15 in stroke, with 3 ft 6 in diameter driving wheels.

On the earlier locomotives of this type the centre-pin of the power-bogie was held by a cast-iron saddle riveted to the boiler, and the tanks and cab were carried on a separate frame bolted to the sides of the firebox. This was satisfactory for the lighter locomotives, but later the centre-pin was attached to a main frame extending nearly the whole length of the locomotive.

The steam pipe was carried through the front tube-plate in the normal manner, then went down to the bottom of the smokebox where it was attached to a horizontal pipe which led back to a hollow bed-

Fig 66 *The* Janus, *the only double-boiler Fairlie-type locomotive designed by William Mason. Built at the Mason Machine Works in 1871. Used on the Lehigh Valley Railroad*

THE MASON-FAIRLIE LOCOMOTIVE

Fig 67 *The* Onward *of the American Fork Railroad; 0–4–4 type. Built at the Mason Machine Works in 1871. 3 ft gauge. This was the first Mason-Fairlie built*

plate casting, in the centre of which was a vertical pipe connected with a ball-and-socket joint and stuffing box. From here a pipe went forward to connect with the branch pipes to the cylinders—thus great flexibility was obtained with only one movable joint. The exhaust pipe had to move with the truck, and the petticoat pipe was made oblong at the lower end to take care of the traverse movements of the blast pipe.

The first locomotives were equipped with the Stephenson valve gear, but in later designs, on account of the closeness of the first pair of driving wheels,

Fig 68 *Locomotive of the 0–4–4 type, built by the Mason Machine Works in 1876. Sold to New York & Manhattan Beach Railroad. 3 ft gauge*

THE FAIRLIE LOCOMOTIVE

Fig 69 *Locomotive of the 0–4–6 type, built by the Mason Machine Works in 1874 for the Erie Railroad*

Fig 70 *The* Taunton *of the 0–6–4 type, built by the Mason Machine Works for stock*

Fig 71 *The* William Mason *of the 0–6–6 type, built by the Mason Machine Works in 1874 for the New Bedford Railroad*

Fig 72 *Locomotive of the 2–4–4 type, built by the Mason Machine Works in 1879 for the Chicago & West Michigan Railroad*

Fig 73 *The Gravesend of the 2–4–6 type, built by the Mason Machine Works in 1881 for the New York & Manhattan Beach Railroad. 3 ft gauge*

Fig 74 *The Breckenridge of the 2–6–6 type, built for the 3 ft gauge Denver, South Park & Pacific Railroad in 1879 by the Mason Machine Works*

THE FAIRLIE LOCOMOTIVE

the Walschaerts valve gear was adopted in 1874, and William Mason was the first American locomotive builder to use this gear. These locomotives were made in all sizes and weights and in eight wheel arrangements. The smallest had cylinders 10 in diameter by 15 in stroke; wheels 3 ft diameter; weight 20 tons, and were for a gauge of 3 ft. Some of the last built had cylinders 18 in diameter by 26 in stroke; wheels 4 ft 6 in diameter, and a total weight of 66 tons. Those first constructed were without pony trucks and were of the 0–4–4, 0–4–6, 0–6–4 and 0–6–6 types.

The 0–4–4 locomotive (Fig 68) with the Mason works insignia on the bunker sides, was built in 1876 (Works No 571) and was used on the 3 ft-gauge West End Passenger Railway constructed in the grounds of the Centennial Exposition, Philadelphia. In January 1877 it was sold to the New York & Manhattan Beach Railway and named *C. L. Flint*. The cylinders were 11 in diameter by 16 in stroke, and the driving wheels had a diameter of 3 ft 6 in.

The 0–4–6 illustrated in Fig 69 was built for the Erie Railroad in 1874 (Works No 528) and was the only Mason-Fairlie on that line, while Fig 70 illustrates an example of the 0–6–4 type, the *Taunton*, which was built for stock, but as the works number is unknown it cannot be identified in the Mason list, and it is not known to what line it was eventually sold.

The *William Mason* (Fig 71) was a 0–6–6 built in 1874 (Works No 536) for the New Bedford Railroad in Massachusetts, and is historically important in that it was the first locomotive to be constructed in America with the Walschaerts valve gear.

William Mason originally intended these locomotives to be run bunker first, but the railroads did not always do this and used them to run either way. With the addition of the pony truck, the 2–4–4, 2–4–6, 2–6–6 and 2–8–6 wheel arrangements were built, and the next four photographs show this development.

The 2–4–4 type (Fig 72) is represented by one of a pair built in 1879 for the Chicago & West Michigan Railroad. The *Gravesend* (Fig 73) was a 2–4–6 built in 1881 (Works No 651) for the 3 ft-gauge New York & Manhattan Beach Railroad, a line that operated seventeen Mason-Fairlies. The next photograph (Fig 74) shows the locomotive *Breckenridge* of the far-famed Denver, South Park & Pacific Railroad in Colorado. This was built in 1879 (Works No 612) and illustrates the 2–6–6 type as constructed for the 3 ft gauge. This railroad was another extensive user of these locomotives, and puchased twenty-three between 1878 and 1880. The ultimate development of the Mason-Fairlie was the 2–8–6 wheel arrangement, and the photograph of the *Denver* (Fig 75), built in 1880 (Works No 632) for the Denver, South Park & Pacific Railroad, shows how impressive in appearance this design could be on a gauge of 3 ft.

All the above photographs (Figs 66 and 68-75) are taken from the large official Mason glass negatives which are now preserved at Silver Spring, Maryland.

From the appended list of railroads that purchased this type of articulated locomotive it will be seen that some 148 were constructed at the Mason Machine Works between 1871 and 1889.

Fig 75 *The final development of the Mason–Fairlie; the* Denver *of the 2–8–6 type, built by the Mason Machine Works in 1880 for the Denver, South Park & Pacific Railroad. 3 ft gauge*

SINGLE BOILER FAIRLIE TYPE LOCOMOTIVES CONSTRUCTED AT THE MASON MACHINE WORKS, 1871–1889

Railroad	State or Country	Name	No	Type	Works No	Date	Cylinders (inches)	Driving Wheels ft in	Weight (lb)	Gauge ft in	Later Operating Company
Allouez Mining Co	Michigan	Gratiot	2	4-coupled	698	1883	14 × 16	3 0			
American Fork Railroad	Utah	American Fork	1	0-4-4	461	1871	10 × 15	" 1		3 0	
Atlantic, Mississippi & Ohio RR	Virginia		86	"	695	1883	14 × 22	4 1		4 8½	Norfolk & Western
"	"		87	"	696	1883	"	"		"	"
Bethlehem Iron Co	Pennsylvania	Kraft	12	6-coupled	667	1881	17 × 24	" 2	77,700		
Boston & Maine RR	Massachusetts	Mathew Craddock	100	2-4-4	746	1887	15 × 22	4 2		4 8½	
Boston, Revere Beach & Lynn RR	"	Orion	1	0-4-4		1875	10 × 16	3 6		3 0	
"	"	Jupiter	3	"	550	1875	"	"		"	
"	"	Pegasus	2	"	549	1876	"	"		"	
"	"	Draco	6	"	559	1876	"	"		"	
"	"		4	2-4-4	683	1882	14 × 18	4 0		"	
"	"		7	"	684	1882	"	"		"	
"	"		8	"	692	1882	"	"		"	
"	"		5	"	720	1885	"	4 1		"	
"	"		6	"	727	1885	"	"		"	
"	"		9	"	740	1887	"	"		"	
"	"		10	"	741	1887	"	"		"	
Burlington & Lamoille RR	Vermont	Lamoille		0-6-6	580	1877	15 × 22	4 0	38,150		Central Vermont
"	"	Mansfield		"	586	1877	"	"			"
Calumet & Hecla Mining Co	Michigan	Calumet		0-6-4	457	1872	16 × 22	3 0		4 0	
"	"	Red Jacket	4	"	622	1880	"	"		"	
"	"	Raymbault		"	681	1882	"	3 6		"	
"	"	Schoolcraft		"	747	1887	16 × 24	4 6	62,800	"	
"	"	St Louis		"	748	1887	"	"		"	
Central Iowa RR	Iowa	Giles E. Taintor	27	0-6-6	655	1881	17 × 24	4 1	48,000		Minneapolis & St Louis
"	"	H. I. Boardman	28	"	656	1881	"	"			"
"	"	Geo. T. M. Davis	29	"	660	1881	"	"			"
"	"	Peter Stan	30	"	661	1881	"	"			"
"	"	Henry L. Jones	31	"	662	1881	"	"			"
"	"	A. L. Burdett	32	"	666	1881	"	"			"
Chicago & Michigan Lake Shore RR	Michigan		8	2-4-4	598	1879	12 × 16	"	31,880		Pere Marquette
Chicago & West Michigan RR	"			"	605	1879	13 × 16	"			"
"	"			"	606	1879	"	"			"
Cincinnatti Northern RR	Ohio	Dakota	4	6-coupled	645	1881	12 × 16	3 0			New York Central
Covington, Columbus & Black Hills	"		1	"	560	1876	"	2 9			Chicago & North Western
Denver & New Orleans RR	Colorado	Cherry Creek	10	0-6-6	673	1882	17 × 24	4 1	77,700	4 8½	Colorado & Southern
"	"	Platte	11	"	674	1882	"	"		"	"

SINGLE BOILER FAIRLIE TYPE LOCOMOTIVES CONSTRUCTED AT THE MASON MACHINE WORKS, 1871–1889

Railroad	State or Country	Name	No	Type	Works No	Date	Cylinders (inches)	Driving Wheels ft in	Weight (lb)	Gauge ft in	Later Operating Company
Denver, South Park & Pacific RR	Colorado	Ore City	3	2–6–6	591	1878	13 × 16	3 1	43,850	3 0	Union Pacific/Chicago, Burlington & Quincy
,,	,,	San Juan	4	,,	597	1878	,,	,,	,,	,,	,,
,,	,,	Ten Mile	6	,,	599	1879	,,	,,	,,	,,	,,
,,	,,	Gunnison	7	,,	600	1879	,,	,,	,,	,,	,,
,,	,,	Lake City	8	,,	601	1879	,,	,,	,,	,,	,,
,,	,,	Kenosha	9	,,	602	1879	14 × 16	,,	45,000	,,	,,
,,	,,	Granite	10	,,	607	1879	13 × 16	,,	43,850	,,	,,
,,	,,	Ouray	11	,,	608	1879	,,	,,	,,	,,	,,
,,	,,	Como	12	,,	609	1879	,,	,,	,,	,,	,,
,,	,,	Ruby	13	,,	610	1879	,,	,,	,,	,,	,,
,,	,,	Twin Lakes	14	,,	611	1879	,,	,,	,,	,,	,,
,,	,,	Breckenridge	15	,,	612	1879	,,	,,	,,	,,	,,
,,	,,	Eureka	16	,,	613	1879	,,	,,	,,	,,	,,
,,	,,	Silverton	20	,,	614	1879	,,	,,	,,	,,	,,
,,	,,	Pitkin City	21	,,	615	1879	,,	,,	,,	,,	,,
,,	,,	Crested Butte	22	,,	616	1879	,,	,,	,,	,,	,,
,,	,,	Grant	23	,,	617	1880	,,	,,	,,	,,	,,
,,	,,	Buena Vista	24	,,	618	1880	,,	,,	,,	,,	,,
,,	,,	Alpine	25	2–8–6	623	1880	15 × 20	3 0	55,340	,,	,,
,,	,,	Rico	26	,,	624	1880	,,	,,	,,	,,	,,
,,	,,	Roaring Fork	27	,,	628	1880	,,	,,	,,	,,	,,
,,	,,	Denver	28	,,	632	1880	,,	,,	,,	,,	,,
Denver, Utah & Pacific RR	,,	Middle Park	11	,,	670	1882	15 × 18	,,	50,610	,,	Chicago, Burlington & Quincy
,,	,,	Denver	10	,,	691	1882	,,	,,	,,	,,	,,
Detroit, Hillsdale & South Western RR	Michigan		5		621	1880					New York Central
Erie Railroad	New York	A.W. Minor	499	0–4–6	528	1874	15 × 22	4 0	61,600	4 8½	,,
Friendship Railroad	,,		1	4-coupled	652	1881	10 × 16	3 6			Pittsburg, Shawmut & Northern
Galveston, Harrisburg & San Antonio RR	Texas	Dixie Crosby	22	0–6–6	563	1876	15 × 20	3 0			Southern Pacific
,,	,,	Com. Garrison	26	6-coupled	570	1876	16 × 24	4 1			
Hecla & Torch Lake RR	Michigan	Torch Lake		0–6–4	518	1873	16 × 22	3 0		4 0	
Illinois Midland RR	Illinois	A.G. Hunter			554	1875	16 × 24	3 6	73,920		Pennsylvania
Juragua Iron Company	Cuba	Juraguacito	1	6-coupled	708	1883	15 × 18	3 0	50,610		
,,	,,	La Folie	2	,,	712	1884	,,	,,	,,		
Long Island Railroad	New York		56	2–4–6	699	1883	14 × 16	4 0	40,000	4 8½	
,,	,,		57	,,	700	1883	,,	,,	,,		
,,	,,		58	,,	701	1883	,,	,,	,,		
,,	,,		59	,,	702	1883	,,	,,	,,		

SINGLE BOILER FAIRLIE TYPE LOCOMOTIVES CONSTRUCTED AT THE MASON MACHINE WORKS, 1871–1889

Railroad	State or Country	No	Name	Type	Works No	Date	Cylinders (inches)	Driving Wheels ft in	Weight (lb)	Gauge ft in	Later Operating Company
Long Island Railroad	New York	60	*East End*	2-4-6	703	1883	14 × 16	4 0	40,000	4 8½	
Marine Railroad	"		*West End*	2-4-4	603	1879	10 × 16	3 1		3 0	
"	"			"	604	1879	"	"		"	
FC Central De Mexico	Mexico	124		2-6-6	752	1889	16 × 24	4 6	93,250	"	National of Mexico
"	"	125		"	753	1889	"	"		"	"
Nantasket Beach Railroad	Massachusetts	23	*Grover Cleveland*	2-4-4	723	1885	14 × 18	4 1	44,000	"	New York, New Haven & Hartford
New Bedford Railroad	"	5	*Wm. Mason*	0-4-4	536	1874	16 × 24	3 6	73,920	"	
New Brunswick Railroad	Canada	6		"	526	1874	12 × 16	3 0		3 6	Canadian Pacific
"	"	7		"	527	1874	"	"		"	"
"	"	8		"	531	1874	"	3 6		"	"
"	"			"	532	1874	"	"		"	"
New York & Brighton Beach RR	New York	1		4-coupled	625	1880	14 × 18	4 0		"	Long Island
"	"	2		"	626	1880	14 × 16	3 0		"	Long Island
New York & Manhattan Beach RR			*C. L. Flint*	0-4-4	571	1877	11 × 16	3 0		3 0	
"			*Admiral Almy*	"	581	1877	12 × 16	3 6		"	
"			*Manhattan* (i)	"	582	1877	"	"		"	
"			*New York* (i)	"	585	1877	"	"		"	
"			*Bay Ridge*	"	588	1877	"	"		"	
"			*Peter Styvesant*	"	590	1878	"	"		"	
"			*Wouter Von Twiller*	"	592	1878	"	"		"	
"			*Washington Irving*	"	593	1878	"	"		"	
"			*Green Point*	"	594	1878	"	"		"	
"			*Brooklyn*	"	595	1878	"	"		"	
"			*Hendrick Hudson*	"	596	1878	"	"		"	
"			*Manhattan* (ii)		648	1881	14 × 18	4 0		"	
"			*Wm. Keift*		649	1881	"	"		"	
"			*East New York*		650	1881	"	"		"	
"			*Gravesend*	2-4-6	651	1881	"	"		"	
"			*New York* (ii)		682	1882	"	"		"	
"			*Oriental*		685	1882	12 × 16	3 6		"	
North Pacific Coast RR	California	2	*San Rafael*	0-4-4	537	1874	13 × 16	3 1		"	North Western Pacific
"	"	8	*Bully Boy*	0-6-6	584	1877	16 × 24	4 9	72,000	"	
Old Colony Railroad	Massachusetts	1		4-coupled	715	1885	10 × 15	3 0		"	New York, New Haven & Hartford
"	"	19		"	721	1885	"	"		"	"
North & South of Georgia RR	Georgia	2		"	508	1873	12 × 16	2 10	42,560	3 0	
Peach Bottom Railroad	Pennsylvania	3		6-coupled	561	1876	12 × 16	4 8½	61,800	4 8½	
Providence, Warren & Bristol RR	Rhode Island	7	*Pokanoket*	2-4-6	726	1885	16 × 24	4			

SINGLE BOILER FAIRLIE TYPE LOCOMOTIVES CONSTRUCTED AT THE MASON MACHINE WORKS, 1871–1889

Railroad	State or Country	Name	No	Type	Works No	Date	Cylinders (inches)	Driving Wheels ft in	Weight (lb)	Gauge ft in	Later Operating Company
Rivière Du Loup Railroad	New Brunswick		2	4-coupled	487	1873	10 × 15	2 9		3 0	Canadian National
"	"		3	"	509	1873	12 × 16	3 0		"	"
St Joseph Lead Company	Missouri		4	"	510	1873	"	"		"	
South Atlantic & Ohio RR	Virg. and Tenness.	Jas. L. Hathaway		2-6-6	643	1881	15 × 22	4 2		4 8½	Southern
"	"		3	"	749	1887	16 × 24	4 6		"	"
Kansas Central Railroad	Kansas	L. T. Smith		6-coupled	750	1887	12 × 16	2 10		3 0	
South Florida Railroad	Florida	R. M. Pulzifer	5	2-6-6	589	1878	13 × 16	3 7	33,600	"	Atlantic Coast Line
"	"	H. B. Plant		4-coupled	704	1883	"	"		"	"
"	"	H. S. Hines		"	705	1883	"	"		"	"
Stockton & Ione Railroad	California	Stockton	9	0-6-4	707	1883	12 × 16	2 9		"	Southern Pacific
"	"	Amador	1	"	551	1875	"	"		"	"
Toledo, Delphos & Burlington RR	Ohio	Geo. W. Kneisley	2	"	552	1875	"	"		"	
"	"	L. Downing Jr	12	6-coupled	627	1880	13 × 16	3 0		"	New York, Chicago & St Louis
"	"	Geo. Wm. Ballou	13	"	629	1880	"	"		"	"
"	"	Calvin S. Brice	14	"	630	1880	"	"		"	"
"	"	Jas. Irvin	15	"	631	1880	"	"		"	"
"	"	G. W. T. Riley	16	6-coupled	635	1880	"	"		"	"
Toledo, Wabash & Western RR	Ohio and Indiana	Hero	17	"	636	1880	"	"		"	"
Utica, Ithaca & Elmira RR	New York	Shoo Fly	187	0-4-4	525	1874	15 × 22	4 0	61,600	4 8½	Wabash
"	"	Leviathan		0-6-6	506	1873	12 × 16	2 9		"	Lehigh Valley
Wheeling & Lake Erie RR	Ohio			4-coupled	547	1875	17 × 24	3 6	73,920	"	New York, Chicago & St Louis
"	"		1	"	646	1881	15 × 22	"		"	"
"	"		2	0-5-6	647	1881	17 × 24	4 1	77,700	"	"
"	"		3	"	653	1881	"	"		"	"
"	"		4	"	654	1881	"	"		"	"
"	"		5	6-coupled	659	1881	15 × 22	3 6	48,000	"	"
"	"		8	4-coupled	668	1882	"	"		"	"
"	"		9	"	669	1882	"	"		"	"
"	"		10	0-6-6	671	1882	17 × 24	4 1	77,700	"	"
"	"		11	"	672	1882	"	"		"	"
"	"		12	"	675	1882	"	"		"	"
"	"		13	"	676	1882	"	"		"	"
"	"		14	"	677	1882	"	"		"	"
"	"		15	"	678	1882	"	"		"	"
"	"		16	"	679	1882	"	"		"	"
"	"		17	"	680	1882	"	"		"	"
"	"		18	4-coupled	686	1882	15 × 22	3 6		"	"

THE MASON-FAIRLIE LOCOMOTIVE

A number of railroads already using Mason-Fairlies purchased others secondhand and these were the Long Island, who obtained one from the New York & Manhattan Beach; the Old Colony who bought Works No 750 from the South Atlantic & Ohio; the Covington, Columbus & Black Hills obtained the *Dixie Crosby* from the Galveston, Harrisburg & San Antonio. The Kansas Central No 5 *L. T. Smith* was sold in 1878 to the Denver, South Park & Pacific, becoming No 5 *Leadville*.

Other railroads used secondhand Mason-Fairlies only, and of these the Colombia & Puget Sound bought Stockton & Ione No 2 and named it *A. A. Denny*; the Cleveland & Marietta bought Wheeling & Lake Erie No 17 and numbered it 11; the Denver, Texas & Fort Worth bought Denver & New Orleans Nos 10 and 11; the Burlington & North Western bought Denver, Utah & Pacific No 10 and numbered it 11; the Herkimer, Newport & Poland bought one 2-4-4 from the New Brunswick and numbered it 2; while the Hoosac Tunnel & Wilmington bought one 0-4-4 from the New Brunswick, while the Boston, Clinton & Fitchburg bought the *William Mason* from the New Bedford.

Other users of Mason-Fairlie locomotives were the Howell & Aspinwall Railroad which according to *Engineering* for 1874 possessed two 0-4-4s at that date, while the South Side Railroad of Virginia was reported by the *Railroad Gazette* for 2 March 1873 as having two. The Central Railroad of Minnesota and the Virginia & Tennessee Railroad both possessed one each at some period. Whether the above six locomotives were all secondhand is not known, but the appended Mason Works list is considered by the leading American authorities to be complete.

Perhaps the most interesting application of the Mason-Fairlie locomotive was on the standard-gauge Mexican Central Railway, a line with many sharp curves and heavy gradients, and which was originally constructed almost without earthworks or ballast. Due to this inferior construction it was necessary to employ articulated locomotives. This railway forms the Mexican section of the direct line from Denver to Mexico City via El Paso, and was united with the American section in 1884. The distance from El Paso to Mexico City is 1,225 miles. Two Mason-Fairlies of the 2-6-6 type were tried on passenger service, and the photograph (Fig 76) illustrates locomotive No 124, built in 1889, and gives a good

Fig 76 *Mexican Central Railway. View of 2-6-6 type Mason–Fairlie No 124, built in 1889 by the Mason Machine Works, on main-line service in difficult country*

THE FAIRLIE LOCOMOTIVE

idea of the terrain traversed by the line. However, these locomotives did not come up to the expectations which had been entertained in using them on arduous main-line service, and the type was discontinued. As a matter of historical interest it may be mentioned here that the three Johnstone 2–6–6–2 double-boiler locomotives built for this line in 1892 were no more successful, and thereafter Mallet locomotives were used.

One railroad on which these Mason-Fairlies were used exclusively was the Boston, Revere Beach & Lynn, a 3 ft-gauge line carrying passengers only, that ran from East Boston to Lynn, about twelve miles, and around the Winthrop Loop. The Mason Machine Works ceased to build locomotives in 1890, and the last two Mason-Fairlies were delivered to the BRB & LRR in 1887, road Nos 9 and 10. But this railroad insisted on having Mason-Fairlie locomotives and no other type, and so the patterns and drawings were handed over first to the Taunton Locomotive Manufacturing Co of Taunton, Massachusetts, who built two 2–4–4 locomotives, road Nos 11 and 12 (Fig 77). The Taunton company went out of business in 1890 and no further Mason-Fairlies were required by the railroad until 1900 when an order was placed with the Manchester Locomotive Works of Manchester, New Hampshire.

This firm was amalgamated in 1901 with seven other major builders to form the American Locomotive Co, and further orders continued to be placed with the Manchester Works.

By 1912, twelve more 2–4–4 locomotives had been built for the BRB & LRR, road Nos 8 and 13-23. No 8 of 1900 is illustrated in Fig 78. In 1914 the railroad ordered three more, Nos 24-26, from the Schenectady Works of the American Locomotive Co, and the photograph (Fig 79) illustrates one of these handsome machines.

With each order the railroad company insisted on the same type of locomotive as furnished by the Mason Works, and they were duplicates in almost every detail. It may be noticed that the 1914 series had heavier balance weights to the driving wheels. Thus the patterns and drawings served three builders in addition to the original firm, and almost twice as many more locomotives of this type were built by these concerns for the Boston, Revere Beach & Lynn Railroad as came from the Mason Machine Works.

For thirty-nine years this design of locomotive was perpetuated on this railroad and rendered most efficient service; several were rebuilt and reboilered, and nearly all were in service when the line was electrified in the 1930s.

Fig 77 *Boston, Revere Beach & Lynn Railroad. Mason–Fairlie type 2–4–4 locomotive No 12, built in 1890 by the Taunton Locomotive Manufacturing Co. 3 ft gauge*

THE MASON-FAIRLIE LOCOMOTIVE

Fig 78 Boston, Revere Beach & Lynn Railroad. Mason–Fairlie type 2–4–4 locomotive No 8, built in 1900 by the Manchester Locomotive Works. 3 ft gauge

Fig 79 Boston, Revere Beach & Lynn Railroad. Mason–Fairlie type 2–4–4 locomotive No 24, built in 1914 at the Schenectady Works of the American Locomotive Co. 3 ft gauge

CHAPTER 19

THE PÉCHOT-BOURDON LOCOMOTIVE 1888-1921

THE Péchot-Bourdon type of double-boiler locomotive, which did not differ materially from a true Fairlie, was constructed according to the combined ideas of Péchot, a captain of artillery, and Bourdon, who was a civil engineer. The design was patented on 3 June 1887, and the following four modifications were claimed: 'A central position for the steam dome ensuring that the steam shall always be taken at a constant height above the water level of the boiler; the use of spring compensation between the boiler and the rear of the bogies in order better to equalise the loads on the axles; simplification of the steam and exhaust piping; attachment of the draw-gear near the centres of the bogies' (Fig 80).

The twin boiler rested on a plate frame forming a cradle for the firebox. These frames were bolted to two saddles, which supported the boiler barrels but were not fixed to them, thus allowing the latter

Fig 80 *General arrangement drawing of Péchot–Bourdon locomotive of 1888, for the French Military Railways. 1 ft 11⅝ in gauge*

THE PÉCHOT-BOURDON LOCOMOTIVE

Fig 81 *Péchot–Bourdon locomotive in service on the Chemin de fer Militaire stratégique, Belfort. 1 ft 11⅝ in gauge*

free movement for expansion, while the frame took all traction stresses. The two power-bogies, each built up of steel plate frames outside the wheels, were attached to the saddles by central pivots through which the steam supply pipes passed so that the joints of the pipes coincided with the bogie centres. The cylinders being under the smokeboxes, the exhaust steam connections were very simple, a ball-jointed pipe giving sufficient flexibility.

These locomotives were first employed on the French narrow-gauge strategic railways of the Toûl fortified zone, where they were mainly used for artillery transport, even guns of 244 mm calibre being carried on this 60 cm gauge. They were also used at the fortresses at the other frontier towns of Belfort, Epinal and Verdun (Fig 81). Fifty-two such locomotives are known to have been built for this service, of which thirty-nine were built by Etabliments Cail, as follows:

Works Nos 2271 to 2289 in 1888   Road Nos 2 to 20
Works Nos 2377 to 2380 in 1892-3
Works Nos 2794 to 2809 in 1906   Road Nos 41 to 56

Twelve were built by Compagnie des Fives-Lille as follows: Works Nos 2769 to 2780 in 1889 and 1890. Road Nos 21 to 32.

The locomotive illustrated in Fig 82, carrying the number 1 and with the name *France*, may have been the first Péchot-Bourdon type to be built. Judging from the shape of the works plate, the builder was the firm of Decauville Aine of Corbeil. The engraving has been taken from a detailed work on French military railways.

The only attempt to use this type of locomotive for other than military service was at the Paris Exhibition of 1889, when one such locomotive was run experimentally.

These locomotives had cylinders $7\frac{1}{8}$ in diameter by $9\frac{1}{2}$ in stroke, with valves of the circular balanced type and Walschaerts gear; wheels 2 ft $1\frac{5}{8}$ in diameter; bogie wheelbase 2 ft $11\frac{1}{2}$ in; total wheelbase 12 ft 6 in; diameter of boiler barrels 2 ft $3\frac{5}{8}$ in, each containing 96 tubes of $1\frac{3}{4}$ in diameter; total heating surface 236·5 sq ft; firebox 38·7 sq ft; total 275·2 sq ft; grate area 5·4 sq ft; working pressure 170·6 psi; centre line of boiler above rail level 3 ft $4\frac{1}{2}$ in; height from rail to top of chimney 9 ft $10\frac{1}{8}$ in; width 6 ft 5 in; overall length 18 ft 11 in; capacity of coal bunkers 12 cwt; the weight in working order was 12 tons 2 cwt.

With the outbreak of World War I in August 1914, the French government immediately sent a mission to the United States to arrange purchases

THE FAIRLIE LOCOMOTIVE

of essential transportation equipment. Orders were placed with the Baldwin Locomotive Works for over 1,000 steam and gasoline locomotives for both the standard and narrow gauges, and among these were 280 locomotives of the Péchot-Bourdon type, suitable for the 1 ft $11\frac{5}{8}$ in-gauge, and intended for use on the tactical light railways to fill the gap between the standard-gauge railways and the front-line positions of the Western Front. These American locomotives were built in a remarkably short space of time, the order being received at Eddystone on 1 February 1915, and by 24 April 1915 the first hundred had been shipped across the Atlantic; each one complete except for the chimneys, and one or two minor fittings, which were packed separately. The remaining 180 of this order were delivered during 1915 and 1916.

These locomotives were classified by Baldwin as 8-8/8. CC. and the individual class numbers, works numbers and running numbers, as far as can now be ascertained, were as follows:

Baldwin Class Nos	Works Nos	Running Nos
1-15	41937-41951	62-76
16-35	41957-41976	77-96
36-50	41979-41993	97-111
51-100	42002-42051	112-161
101-110	43081-43090	175-184
111-115	43194-43198	185-188
116-119	43267-43270	190-193
120	43276	194
121-135	43311-43325	195-209
136	43333	210
137-146	43363-43372	211-220
147-150		
151-158	43439-43446	225-232
159-170	43452-43463	233-244
171-175	43507-43511	245-249
176-180	43522-43526	250-254
181-185	43792-43797	255-259
186-193	43873-43880	260-267
194-200	43952-43958	268-274
201-210	43972-43981	275-284
211-213	43999-44001	285-287
214-218	44112-44116	288-292
219-225	44137-44143	293-299
226-229	44158-44161	300-303
230	44167	304
231		
232	44125	306
233-235	44216-44218	307-309
236-240	44255-44259	310-314
241-250	44315-44324	315-324
251-267	44409-44425	325-341
268-270	44449-44451	342-344
271-279	44464-44472	345-353

Nineteen locomotives, bearing Running Nos 130, 152, 155, 157, 159, 210, 272, 282, 291, 304, 306, 313, 316, 317, 318, 321, 325, 348, 352, were shipped to Algeria when new and not the Western Front. It

Fig 82 *Péchot–Bourdon locomotive* France, *built in 1889 for the French Military Railways. 1 ft $11\frac{5}{8}$ in gauge*

THE PÉCHOT-BOURDON LOCOMOTIVE

Fig 83 *Péchot–Bourdon locomotive built by the Baldwin Locomotive Works in 1921 for the Japanese Government. 1 ft 11⅝ in gauge*

will be observed that this list enumerates 279 locomotives. Fifteen similar locomotives, ordered by the French Government, were constructed at the Atlas Works of the North British Locomotive Co, Glasgow, in 1915 (Works Nos 21185-21197).

All these locomotives, built to metric measurements, had cylinders of 6·88 in diameter by 9·44 in stroke; wheels 2 ft 1⅝ in diameter; bogie wheelbase 2 ft 11½ in; total wheelbase 12 ft 6 in; overall length 18 ft 11 in; tube heating surface 249·99 sq ft; fireboxes 40·491 sq ft; total heating surface 290·481 sq ft; grate area 5·102 sq ft; working pressure 170·6 psi. The two internal copper fireboxes were separated by a 2 in water space. The tanks held 333·23 gallons, and the coal bunkers 7 cwt 98 lb.

Unfortunately, owing to the fact that their smoke disclosed their position, these locomotives could not be used near the advanced positions of the Western Front, and so in this area they were replaced by internal-combustion tractors.

What eventually happened to all these locomotives after 1918 is not clear. The disposal of such a large number of specialised machines of so small a gauge could not have been an easy matter, and probably many were broken up within a few years. Some, however, survived many years, and during World War II two were transferred by the Germans (from some unknown source) to the lignite opencast mine at Stari Kostolac, east of Belgrade, and worked there up to the 1950s until the deposits were exhausted and the site closed. One of these two locomotives has been preserved at the Yugoslav Railway Museum at Belgrade. Another Péchot-Bourdon locomotive is now in the Museum at Dresden, in East Germany.

In 1921, one Péchot-Bourdon locomotive (Fig 83) was built by the Baldwin Locomotive Works to the order of the Japanese Government. This example has been described, by the engineer in charge of its erection, as an extremely beautiful machine, most carefully constructed and finished. It was apparently the intention of the Japanese to copy it for use on their 1 ft 11⅝ in-gauge military railways, but no Japanese-built Péchot-Bourdon locomotives are known. In the comprehensive work *Steam Locomotives in Japan, 1871-1960* (in Japanese), the date of this Baldwin locomotive is given as 1919.

BIBLIOGRAPHY

Articulated Locomotives, by L. Wiener, 1930.
Avonside Engine Company Catalogue.
The Engineer, London.
Engineering, London.
The Locomotive, London.
Die Lokomotive, Vienna.
Trains (May 1961), USA.
Locomotives of the Railways of the Soviet Union, by V. A. Rakov, 1955. In Russian.
Steam Locomotives in Japan, 1871-1960, 1961. In Japanese.
Unpublished Crown-copyright material in the India Office Records.
Die Entwicklung der Lokomotive, Vol 2, by Baurat Dr Ing Metzeltin, 1937.
Stimmen über die Schmalspurigen Eisenbahnen, by Nordling, 1871.
Specielle Eisenbahntechnik, by Heusinger von Waldeck, 1885.
Organ für die Fortschritte des Eisenbahnwesens, 1872.

Enzyklopädie des Eisenbahnwesens, by V. Roll.
The Festiniog Railway, Vols 1 and 2, by J. I. C. Boyd.
Das Eiserne Jahrhundert, 1884.
Gedanklese über die Wichtighut des Fairlieschen Lokomotive Systems, by Bemarteaua, 1872.
Technische Mittheilungen, Heft 3 of 1876. Zurich.
Sächsische Maschinenfabrik Catalogue, Chemnitz.
Weiner Neustadt Catalogue in Library of Vienna Technical University.
Weiner Neustadt Catalogue in the Railway Museum Library, Vienna.
Voie et Material Roulant, by Couche.
Génie Civil, 1918.
History of the Baldwin Locomotive Works, 1923.
Locomotive Engines, by R. F. Fairlie, 1864. (Reprinted 1969.)
Battle of the Gauges Renewed: Railways or No Railways, by R. F. Fairlie, 1872.
Cavalcade of New Zealand Locomotives, by A. N. Palmer & W. W. Stewart, 1957.

ACKNOWLEDGMENTS

It has taken me many years to gather together the information and illustrations contained in this book, and without the help of many persons with kindred interests it would probably never have been completed.

I should particularly like to thank Thomas Norrell of Silver Spring, Maryland, who has gone to great lengths to obtain photographs for me and in suggesting possible sources of information; also my friend, Rheimar Holzinger of Vienna, for directing my attention to the German works on this subject, while correspondence with the late P. C. Dewhurst resulted in a number of obscure points being cleared up, chiefly with regard to South America. For much of the information on Avonside-built Fairlies for India and Burma I am indebted to H. Hughes of Croydon. The locomotive building industry has readily given me information and photographs, and I am grateful to the technical staffs of the firms concerned for patiently dealing with my inquiries.

I am also greatly indebted to the following:

Eng Antonio de Anda, Montevideo.
Anglo-Chilean Nitrate & Railway Co, Antofagasta.
Mr Bradley, CME Western Australian Government Railways.
J. F. Bruton, Fulham.
A. G. Culpeffer-Cooke, Victoria, Australia.
C. H. Dickson, Ilford.
L. Douglas, South African Railways, Johannesburg.
A. E. Durrant, Johannesburg.
C. Hamilton Ellis, Horsham.
C. E. Fisher, Boston, USA.
A. Fussell, New Zealand Government Railways, Wellington.
A. G. W. Garraway, Festiniog Railway, Portmadoc.
J. A. McGavin, New Zealand.
S. H. Pearce Higgins, Worcester.
D. H. Holland, Johannesburg, SA.
Hunslet Engine Company Ltd, Leeds.
J. P. Hugo, general manager, South African Railways, Johannesburg.
Georges Mangin, Villefranche, France.
North British Locomotive Co Ltd, Glasgow.
K. P. Plant, Sheffield.
H. C. Ritchie, Archivist, City of Schenectady, USA.
Secretary of State for Foreign and Commonwealth Affairs.
M. Seymour, Cambridge.
Erik Sundstrom, Bromma, Sweden.
J. R. M. Thomas, Yatton, Somerset.
C. A. Turner, Act CME Queensland Government Railways, Brisbane.
Vulcan Foundry Limited, Newton-le-Willows, Lancashire.
J. H. White, Smithsonian Institute, Washington, USA.
Yorkshire Engine Company Ltd, Sheffield.

SOURCES OF PHOTOGRAPHS AND DRAWINGS

The Author wishes to thank the following individuals, railway companies, locomotive builders, and other organisations for permission to use illustrations:

M. Seymour, Cambridge (*Frontispiece*).
Ian Allan Ltd, Shepperton (Figs 2, 5, 7, 9, 11, 54, 60).
Anglo-Chilian Nitrate Railway (Fig 41).
Antonio de Anda, Penarol Works, Montevideo (Fig 23).
Editors of *Engineering* (Figs 4, 8, 16, 20, 34, 35, 57a).
Editor of *The Engineer* (Fig 9a).
Effingham Wilson, London (Fig 13).
Festiniog Railway (Figs 6, 32).
Reimar Holzinger, Vienna (Fig 56).
D. F. Holland, Johannesburg (Figs 25, 26, 27).
Hunslet Engine Co, Leeds (Figs 21, 31, 65).
Locomotive & General Railway Photographs, Hockley Heath (Figs 19, 57).
Georges Mangin, Villefranche (Figs 80, 82).
Thomas Norrell, Maryland (Figs 14, 17, 18, 22, 33, 55, 66 to 78, 81, 83).

New Zealand Government Railways (Figs 24, 28, 29).
North British Locomotive Co, Glasgow (Figs 10, 58, 59, 63, 64).
R. W. Kidner, The Oakwood Press (Fig 39).
S. H. Pearce Higgins, Worcester (Fig 3).
Queensland Government Railways (Fig 47).
H. G. Ritchie, Archivist, City of Schenectady, USA (Fig 79).
Science Museum, London (Fig 12).
Controller of HM Stationery Office, London (Fig 1).
Swedish Railway Museum, Stockholm (Fig 53).
J. R. M. Thomas, Yatton (Fig 15).
Vulcan Foundry Ltd, Newton-le-Willows (Figs 42 to 46, 48 to 52).
West Australian Government Railways (Fig 30).
Yorkshire Engine Co, Sheffield (Figs 37, 38, 40).

The illustrations Figs 36, 61, 62 are from unknown sources, and although every endeavour has been made, it has not been possible to establish ownership of copyright, and for this omission the Author begs indulgence.

INDEX

Afghan War (Third), 40
Ahrons, E. L. (1866-1926), 59, 75
Algeria, 94
Allan, Alexander (1809-1891);
    link motion, 20, 34, 46, 61, 65, 77
Allen, Horatio (1802-1889), 11
Alle-Verte depot, Brussels, 45
Alto de Junin (Chile), 48
American Locomotive Co, 90, 91
Arizoba-Esperanza section, Mexican Ry, 12, 56
Atlas Works, Manchester, 24-6, 75, 77
Australia, 12, 15, 39, 50
Avonside Engine Co, Bristol, 27-42, 55, 58, 62, 77, 78

Bailey, Hawkins & Co (agents), 33
Baldwin Locomotive Works, 94, 95
Barbet, P. A. (engineers), Ipswich, Australia, 51
Barquisimeto, Venezuela, 57
Beddgelert, 74
Belfort, 93
Belgrade, museum, 95
Bennett, A. R. (1850-1928), 75
Blackheath Colliery, Ipswich, Australia, 52
Boca de Mato, Brazil, 26
Bogies, 77
Boilers, 77
Bolan Pass, 40
Bombay, 40, 41
Boston Lodge Works, 54, 62, 63, 64
British Government, 74
Briton Ferry Works, 15
Bryngwyn branch (NWNG Ry), 52
Burma, 12, 41, 78
Burry Port, 16, 18

Cabs, 78
Cail & Cie, 93
Canada, 12, 28, 76
Cape Colony, 12
Chile, 20, 28, 48
Chilian Government, 34
Cirencester Workshops (SM & A Ry), 59
Clark, Punchard & Co (contractors), 18
Cockerill Works, Seraing, 11
Colorado, 12, 50
Coonoor (South Indian Ry), 42

Cordova & Bocca del Monte section, Mexican Ry, 12
Cross, James & Co, St Helen's, 13-17, 75
Crossheads, 78
Crown Agents for the Colonies, 34
Cuba, 12, 32

Davies & Metcalfe Ltd, Manchester, 54
Ddualt Station, Festiniog Ry, 64
Decauville Aine, Corbeil, 93, 94
Denver City, 50
Dinas Junction (NWNG Ry), 52
Dresden, museum, 95
Dunedin and South Seas Exhibition, 49

East Boston (USA), 90
Eddystone Plant (Baldwin), 94, 95
England, 12
England, George & Co, Hatcham Ironworks, 11, 14, 16
England, George (junior), 16
Epinal, 93
Exhaust piping, 14, 15, 20, 22, 34, 36, 48, 54, 55, 65, 66, 74, 78

Fairlie Engine & Rolling Stock Co, 21
Fairlie Engine & Steam Carriage Co, 16-20
Fairlie, Robert Francis (1831-85), 11, 12, 16, 21, 38, 43, 46, 75, 76
Fell centre-rail system, 26
Fenton, W. (of Rochdale), 44
Fives-Lille, 93
Frames, 77
France, 12
Fraser, J. S. (of GW Ry), 16
French Government, 93

Gauges (other than 4ft $8\frac{1}{2}$ in):
    1 ft $11\frac{1}{2}$ in   17, 29-31, 52-5, 62, 64, 73
    1 ft $11\frac{5}{8}$ in   93-5
    2 ft 0 in   57, 58
    2 ft $5\frac{1}{2}$ in   59, 69, 70
    2 ft 6 in   48-50
    2 ft $11\frac{1}{2}$ in   54
    3 ft 0 in   29, 32, 50, 76, 85-8
    3 ft $3\frac{3}{8}$ in   40-2, 55

THE FAIRLIE LOCOMOTIVE

    3 ft 6 in   15, 16, 24, 25, 28, 33, 34, 37-9, 48-50, 87
    3 ft $7\frac{5}{16}$ in   25, 26, 44
    3 ft $7\frac{5}{8}$ in   26
    4 ft 0 in   85, 86
    5 ft 0 in   25, 28, 38, 45, 60, 61, 67, 68
    5 ft 3 in   22
Georgia, Russia, 28, 79
Geralton, Western Australia, 39
Ghat incline, Burma State Rys, 56
Ghat inclines, Great Indian Peninsular Ry, 11
Ghat incline, Holkar State Ry, 40
Gooch stationary link motion, 17, 77
Grange Colliery branch, 43

Harnai Valley, India, 40
Hartman, Richard, Chemnitz, Saxony, 69, 70
Hatcham Ironworks, 11, 20, 21, 27, 29, 77
Havana, Cuba, 32
Hawthorn, R. & W. & Co, Newcastle, 57-9
Hendon, 14
Hirok-Kotal section (North Western State Ry, India), 41

Illimoff, M. (of St Petersburg), 43
Inchicore Works, Dublin, 22, 23
India, 12
Indwe (Colonial coal from), 36
Ipswich, Australia, 51
Iquique City, Chile, 20, 79
Ireland, 12, 22

Japan, 56, 95
Japanese Government, 95
Johnstone type locomotives, 90
Joy valve gear, 56, 78
Junin City, Chile, 48

Kentish Town, 14
Khandwa (Holkar State Ry), 40
Kitson & Co, Leeds, 36
Kolomensky Works, Russia, 67, 68

Lahore, 42
Landeburg, M. (of Sweden), 43
La Veta Pass (Colorado), 50
Lemonius & Co, Liverpool (agents), 50
Llantrisant Tinplate Works, 15
Luxembourg, 12
Lynn (Boston, USA), 90

Machinery Registration Depot, Newport, Monmouthshire, 58

Manchester Locomotive Works, Manchester, New Hampshire, 90
Mandalay-Kulon section, Burma State Rys, 41
Maria Elena (nitrate plant), Chile, 48
Mason-Fairlie type locomotives, 80-4, 89-91
Mason Machine Works, Taunton, Massachusetts, 80-9
Mason, William (of American), 21, 80
McDonnell, Alexander (1829-1904), 22
Mexico, 12
Mexico City, 50
Meyer type locomotives, 59, 70
Ministry of Supply, 74
Modified-Fairlie type, 48, 55
Molteno (colonial coal from), 36
Montero Brothers (operators of Iquique Ry), 20
Montevideo, 34
Mooloolaba (sawmill), Queensland, 51
Morton, J., & Sons (contractors), 18

Neilson & Co, Glasgow, 46, 65, 66, 79
New Plymouth Sheds (NZ Gov Rys), 34
New Zealand, 12
New Zealand, General Government of, 33
Niterói, Brazil, 26
North British Locomotive Co, Glasgow, 71, 72, 95
Northampton, Western Australia, 39
Norway, 12
Norwegian Government, 50
Nova Friburgo, Brazil, 26
Nova Scotia, 29
Novosselsky, Nicholas, His Excellency, 43
Nydquist & Holm, Trollhaten, 44

Oatacamund (South Indian Ry), 42
Otago, 34
Otago Early Settlers Hall, 49
Otago Iron Rolling Mills, 49
Otago Provincial Council, New Zealand, 33

Palmer, General, 50
Paris Exhibition, 1878, 58
Paris Exhibition, 1889, 93
Patent Regulating Gear (G. P. Spooner's), 30, 32, 33
Péchot-Bourdon type locomotives, 15, 92-5
Pedro de Valdivia (nitrate plant), 48
Peñarol Works, Montevideo, 34
Performance of locomotives, 78, 79
Peru, 18, 27, 28, 36, 49
Phipps & Co (agents), 25
Pivots (bogie), 13, 15, 17, 20, 22, 25, 30, 40, 53, 62, 65, 73
Portugal, 12, 40, 54
Public Works Department, New Zealand, 49

INDEX

Quartier Leopold depot, Brussels, 45
Queensland, 15, 50, 51, 77

Railway and Locomotive Historical Society, America, 76
Railways:
    Allouez Mining, 85
    American Fork, 80, 81, 85
    Anglesey Central, 14, 15, 17
    Anglo-Chilian Nitrate, 47, 48
    Atlantic Coast Line, 88
    Atlantic, Mississippi & Ohio, 85
    Bethlehem Iron Co, 85
    Bolivar, 57
    Bombay & Baroda, 11
    Boston & Albany, 80
    Boston, Clinton & Fitchburg, 89
    Boston & Maine, 85
    Boston, Revere Beach & Lynn, 85, 90, 91
    Brecon & Merthyr, 14
    Burlington & Lamoille, 85
    Burlington & North Western, 89
    Burma State, 41, 42, 55, 56
    Burry Port & Gwendraeth Valley, 16, 18, 19
    Calumet & Hecla Mining, 85
    Canadian National, 88
    Canadian Pacific, 28, 87
    Canta-Gallo, 25, 26, 44
    Cape Government, 34, 36, 37, 38
    Central Iowa, 85
    Central Pacific, 80
    Central Railroad of Minnesota, 89
    Central Uruguay, 16
    Central Vermont, 85
    Chicago, Burlington & Quincy, 86
    Chicago & Michigan Lake Shore, 85
    Chicago & North Western, 85
    Chicago & West Michigan, 83, 85
    Chimbote-Huaraz, 76
    Cincinnatti Northern, 85
    Cleveland & Marietta, 89
    Colorado & Southern, 85
    Columbia & Puget Sound, 89
    Covington, Columbus & Black Hills, 84, 89
    Credit Valley, 76
    Denver & New Orleans, 85, 89
    Denver & Rio Grande, 50, 51
    Denver, Hillsdale & Southwestern, 86
    Denver, South Park & Pacific, 83, 84, 86, 89
    Denver, Texas & Fort Worth, 89
    Denver, Utah & Pacific, 86, 89
    Deutsche Reichsbahn, 70
    Dundee & Port Chalmers, 49
    East & West Junction, 46, 47, 58
    Erie, 82, 86
    Festiniog, 12, 17, 19, 29, 32, 33, 54, 55, 62-4
    Friendship, 86
    Galveston, Harrisburg & San Antonio, 86, 89
    Glasgow & Cape Breton, 29, 32
    Great Indian Peninsular, 11
    Great Russian, 75
    Great Southern & Western, 22, 23
    Great Western, 16, 18
    Hallsberg-Motala-Mjolby, 43, 44
    Hecla & Torch Lake, 86
    Herkimer, Newport & Poland, 89
    Holkar State, 40
    Hoosac Tunnel & Wilmington, 89
    Howell & Aspinwall, 89
    Illinois & Midland, 86
    Imperial Linvy, 24, 25
    Indian State, 40, 41-2
    Iquique, 18, 27, 28, 34, 76
    Japanese Military, 95
    Junin, 48, 71
    Juragua Iron, 86
    Kansas Central, 88, 89
    Lehigh Valley, 80, 88
    Leopoldina, 26
    Lima-Oroya, 76
    Londonderry & Coleraine, 11
    Long Island, 86, 89
    Luxembourg (Grande Compagnie du), 44, 45
    Madras, 42
    Manchester, Sheffield & Lincolnshire, 43
    Marine, 87
    Matanzas, 32
    Mexican, 28, 30-1, 43, 46, 52, 56, 65, 66, 71, 72, 75, 78, 79
    Mexican Central, 87, 89
    Midland, 14
    Minneapolis & St Louis, 85
    Monmouthshire, 14
    Mont Cenis, 26
    Nantasket Beach, 87
    Nassjo-Oscarshamn, 18, 24, 57
    National of Mexico, 87
    Neath & Brecon, 13, 15, 77
    New Bedford, 82, 87, 89
    New Brunswick, 87, 89
    New York & Brighton Beach, 87
    New York Central, 85, 86
    New York, Chicago & St Louis, 88
    New York, New Haven & Hartford, 87
    New York & Manhattan Beach, 81, 83, 87, 89
    New Zealand Government, 33, 34, 36, 38, 39, 40, 79
    Nilgiri Mountain, 42
    Nitrate Railway, 28, 29, 31, 33, 34, 43
    Nitrate Railways, 28, 34, 46, 78, 79
    Norfolk & Western, 85
    North Pacific Coast, 87
    North & South of Georgia, 87
    North Wales Narrow Gauge, 52, 53, 73, 74
    North Western Pacific, 87
    North Western State, 41
    Northern of Montevideo, 34, 35, 78

Norwegian State, 50, 52, 58
Old Colony, 87, 89
Patillos, 49, 50
Peach Bottom, 87
Pennsylvania, 85, 87
Pere Marquette, 86
Pimental & Chicklaya, 50, 51
Pisagua, 27, 28, 34, 76
Pittsburg, Shawmut & Northern, 86
Porto a Pavoa de Varzim, 53, 54
Poti & Tiflis, 25, 28, 29, 38, 45, 60, 67, 79
Providence, Warren & Bristol, 87
Quelbrada Copper, 57
Rajputana-Malwa State, 40
Riviere du Loup, 88
Russian State, 67, 68
St Joseph Lead Co, 88
Saxony State, 39, 59, 69
South Atlantic & Ohio, 88, 89
South Florida, 88
South Indian, 42
South Side of Virginia, 89
Southern (USA), 88
Southern Pacific, 86, 88
Southern & Western of Queensland, 15, 16, 17, 50, 52, 77
Stockton & Ione, 88, 89
Swindon, Marlborough & Andover, 58, 59
Tamboff & Saratoff, 25, 38, 67
Tarapaca Junction, 76
Toledo, Delphos & Burlington, 88
Toledo, Wabash & Western, 88
Toronto, Grey & Bruce, 28, 76
Toronto & Nipissing, 28
Transcaucasian, 28, 45, 67
Union Pacific, 86
Utica, Ithaca & Elmira, 88
Vendee, 18
Venezuelan Central, 75
Venezuelan Government, 33, 34
Virginia & Tennessee, 89
Wabash, 88
Welsh Highland, 74
West Australian Government, 39
West Australian Land Co, 39, 41
Wheeling & Lake Erie, 88, 89
Reducto, Chile, 48
Reefton Nursery Play Centre (South Island, NZ), 39
Reichenbach, Saxony, 70
Riding characteristics of Fairlie locomotives, 78, 79
Ridley & Young, Darlington, 74
Rio de Janeiro, 26, 44
Roberts, Edmund, 36
Rous-Martin, Charles, 79
Russia, 12, 24, 25, 28, 38, 45, 56, 60, 61, 67, 68, 75

Salitrero (Administration of the Ferrocarril), 34, 46
St Helen's Railway Workshops, 13
Santa Fe, 50
Santa Lucia, Uruguay, 34
Saxon Engine Works, Chemnitz, 69, 70
Saxony, 12
Schenectady Works, American Locomotive Co, 90, 91
Semmering Contest, 11
Senhora da Horo-Matosinhos branch, Portugal, 54
Sharp, Stewart & Co, Manchester, 24-6, 75, 77
Sigl, Vienna, 60, 61, 67
Single-Fairlie type, 12, 22, 23, 38, 39, 40, 52, 53, 54, 58, 59, 73, 79, 81-4, 89-91
Souram Pass, 12, 28, 68, 79
South America, 12
South Snowdon, 52
Spain, 32, 33
Spooner, C. E., 22, 29
Spooner, G. P., 29
Spooner's Patent Regulator Gear, 30, 32, 33, 62
Stari Kostolac, Yugoslavia, 95
Stationary link motion (Gooch), 17, 77
Steam piping, 14, 15, 20, 22, 48, 54, 55, 65, 66, 74, 78, 80, 81, 93
Stephenson valve gear, 34, 44, 53, 77
Sukkur, India, 41, 42
Sutherland, Duke of, 43
Sweden, 12, 24, 43, 44, 57
Swindon Town (SM & A Ry Goods Yard), 59
Swindon Transfer (GWR Goods Yard), 59

Talybont incline, Brecon & Merthyr Ry, 34
Taunton Locomotive Manufacturing Co, Mass, 90
Taunton, Massachusetts, 80, 90
Teodoro de Olivera, Brazil, 26
Third Afghan War, 40
Thouvenot (French Patent), 11
Tocopilla, Chile, 48
Toowoomba, Queensland, 51
Toul area, France, 93
Traeth Mawr embankment, Portmadoc, 62
Transcaucasian Ry, Russia, 28, 45, 67
Trondhjem, Norway, 58
Tucacas, Venezuela, 59

Uruguay, 34

Valve gears, 17, 20, 34, 53, 61, 77, 78
Venezuela, 12, 21, 75
Venezuelan Government, 33, 34
Verdun, 93
Volga, 25
Vulcan Foundry, 49-56, 62, 64, 78

INDEX

Wake's Yard, Darlington, 74
Wales, 12
Walschaert's valve gear, 29, 34, 36, 39, 40, 48, 58, 71, 73, 78, 84
Wanganui section, NZ Gov Rys, 34
Western Front (France 1914-1918), 94
West Point Foundry, New York, 11
Wiener-Neustadter, Vienna, 60, 61
Winthrop Loop (Boston, USA), 90

Woodburning, 25, 28, 29, 43, 50, 61, 68, 78
Wooloongabba railway yards, Brisbane, 51

Yorkshire Engine Co, Sheffield, 16, 43-8, 65, 71, 77, 78

Zibingyi Ghat, Burma State Railways, 56